Food Riots, Food Rights and the Politics of Provisions

Thousands of people in dozens of countries took to the streets when world food prices spiked in 2008 and 2011. What does the persistence of popular mobilization around food tell us about the politics of subsistence in an era of integrated food markets and universal human rights? This book interrogates this period of historical rupture in the global system of subsistence, getting behind the headlines and inside the politics of food for people on low incomes.

The half decade of 2007–2012 was a period of intensely volatile food prices as well as unusual levels of popular mobilization, including protests and riots. Detailed case studies are included here from Bangladesh, Cameroon, India, Kenya and Mozambique. The case studies illustrate that political cultures and ways of organizing around food share much across geography and history, indicating common characteristics of the popular politics of provisions under capitalism. However, all politics are ultimately local, and it is demonstrated how the historic fallout of a subsistence crisis depends ultimately on how the actors and institutions articulate, negotiate and reassert their specific claims within the peculiarities of each policy.

A key conclusion of the book is that the politics of provisions remain essential to the right to food and that they involve unruliness. In other words, food riots work. The book explains how and why they continue to do so even in the globalized food system of the 21st century. Food riots signal a state unable to meet a principal condition of its social contract, and create powerful pressure on a state to address that most fundamental of failings.

Naomi Hossain is a political sociologist at the Institute of Development Studies at the University of Sussex, UK. Her most recent book is *The Aid Lab: Understanding Bangladesh's Unexpected Success* (Oxford University Press, 2017).

Patta Scott-Villiers is a political sociologist and convenes the Power and Popular Politics Cluster at the Institute of Development Studies at the University of Sussex, UK. She is lead author of *Precarious Lives: Food Work and Care after the Global Food Crisis* (Institute of Development Studies, 2016).

Routledge Studies in Food, Society and the Environment

For further details please visit the series page on the Routledge website: www.routledge.com/books/series/RSFSE/

Food Consumption in the City
Practices and Patterns in Urban Asia and the Pacific
Edited by Marlyne Sahakian, Czarina Saloma and Suren Erkman

Eating Traditional Food
Politics, Identity and Practices
Edited by Brigitte Sébastia

Feeding Cities
Improving Local Food Access, Security and Sovereignty
Edited by Christopher Bosso

The Right to Food Guidelines, Democracy and Citizen Participation
Country Case Studies
Katharine S.E. Cresswell Riol

Peasants Negotiating a Global Policy Space
La Vía Campesina in the Committee on World Food Security
Ingeborg Gaarde

Public Policies for Food Sovereignty
Social Movements and the State
Edited by Annette Desmarais, Priscilla Claeys and Amy Trauger

Sustainable Food Futures
Multidisciplinary Solutions
Edited by Jessica Duncan and Megan Bailey

Food and Cooking Skills Education
Why Teach People How to Cook?
Anita Tull

Food Riots, Food Rights and the Politics of Food Provision
Edited by Naomi Hossain and Patta Scott-Villiers

Food Riots, Food Rights and the Politics of Provisions

Edited by Naomi Hossain and Patta Scott-Villiers

Routledge
Taylor & Francis Group
LONDON AND NEW YORK
earthscan
from Routledge

First published 2017
by Routledge

2 Park Square, Milton Park, Abingdon, Oxfordshire OX14 4RN
52 Vanderbilt Avenue, New York, NY 10017

Routledge is an imprint of the Taylor & Francis Group, an informa business

First issued in paperback 2019

British Library Cataloguing-in-Publication Data
A catalogue record for this book is available from the British Library

Library of Congress Cataloging-in-Publication Data
Names: Hossain, Naomi, editor. | Scott-Villiers, Patta, editor.
Title: Food riots, food rights and the politics of provisions / edited by Naomi Hossain and Patta Scott-Villiers.
Description: New York : Routledge, 2017. | Series: Routledge studies in food, society and the environment | Includes bibliographical references and index.
Identifiers: LCCN 2017016705 | ISBN 9781138040168 (hardback) | ISBN 9781315175249 (ebook)
Subjects: LCSH: Food riots—History—21st century. | Right to food. | Food supply—Government policy.
Classification: LCC HV6474 .F66 2017 | DDC 363.809/051—dc23
LC record available at https://lccn.loc.gov/2017016705

ISBN: 978-1-138-04016-8 (hbk)
ISBN: 978-0-367-35215-8 (pbk)

Typeset in Goudy
by Apex CoVantage, LLC

Contents

Figures and tables

Figures

Tables

Contributors

Sara Burke is a Senior Policy Analyst in the New York office of Friedrich-Ebert-Stiftung, a political think tank headquartered in Berlin, Germany. Her research focuses on global economic governance and the multilateral system, especially on the impacts economic and social inequalities and political participation at all levels of governance.

Egídio Chaimite is a researcher with the Citizenship and Governance Research Group of the Institute of Social and Economic Research (IESE) in Maputo, Mozambique and a PhD candidate at the Institute of Development Studies, University of Sussex. He works on mobilization and collective action, governance, policy and the state. He has a degree in Public Administration from Universidade Eduardo Mondlane (UEM) and a Masters in Political Science from the Université de Bordeaux.

Luís de Brito is Research Director of the Institute of Social and Economic Research (IESE) and an Associate Professor of the Faculty of Arts and Social Sciences at Universidade Eduardo Mondlane (UEM) in Maputo, Mozambique. A member of IESE's Citizenship and Governance Research Group, he works on elections and political representation. He has a PhD in The Anthropology and Sociology of Politics from the Université Paris VIII.

Naomi Hossain is a political sociologist at the Institute of Development Studies at the University of Sussex, UK. Her most recent book is *The Aid Lab: Understanding Bangladesh's Unexpected Success* (Oxford University Press, 2017).

Ferdous Jahan is Professor of Public Administration at the University of Dhaka, and was previously Academic Coordinator at the BRAC Institute of Governance and Development at BRAC University. Jahan teaches a number of topics on the intersection between development, governance and politics, and has worked on issues of governance, social protection, urban poverty, legal empowerment of the poor and women's empowerment in a number of developing country settings. Her current research interests governance and the judiciary, as well as public policy processes in food security, social protection, urban poverty and gender, on which she has published widely.

Anuradha Joshi is a senior research fellow and leads the Governance Cluster at the Institute of Development Studies at the University of Sussex, UK. She is co-Research Director of the Action for Empowerment and Accountability Research Consortium. The focus of her work is on state-society relations and accountability, mainly in areas of service delivery, taxation of the informal economy and policy processes. She currently serves on several international advisory panels.

Devangana Kalita is an independent researcher and activist based in Delhi, India with a Masters in Gender and Development from the Institute of Development Studies, UK and a Masters in Modern Indian History from Jawaharlal Nehru University, India.

Celestine Nyamu Musembi is a Kenyan lawyer with a background in legal anthropology. She is a Senior Lecturer in law at the University of Nairobi, and a former research fellow at the Institute of Development Studies, Sussex, UK. Her teaching and research interests include human rights and development, gender, culture and property relations, housing rights and accountability of local governance institutions.

Bonface Omondi is an advocate of the High Court of Kenya and Legal Research Consultant.

Biraj Patnaik has recently taken over as the Regional Director for South Asia at Amnesty International. He was the Principal Adviser to the Commissioners of the Supreme Court of India on the right to food for over a decade. Biraj was closely involved with the processes of drafting and lobbying for the National Food Security Act and has been associated with the Right to Food Campaign in India from its inception. He is on the International Editorial Board of the Global Food and Nutrition Watch.

Lucio Posse is a researcher at the Institute of Social and Economic Studies (IESE). He has a Master's degree in African Studies from the University Institute of Lisbon (ISCTE-IUL), Portugal. His research areas are local governance, decentralization, nation, nationalism and identity.

Vaibhav Raaj is a New Delhi-based independent researcher with an MPhil in Political Studies from Jawaharlal Nehru University in India. His research interests include the labour market and political mobilizations of informal labour, gender relations, public policy, agrarian politics, social impacts of climate change and such other diverse topics.

Muhammad Ashikur Rahman is a Research Associate of BRAC Institute of Governance and Development (BIGD) at BRAC University in Dhaka, Bangladesh. He received his Bachelors and Masters degrees in Public Administration from the University of Dhaka. His interests are labour and other forms of social movements, local governance, institutions and electoral systems of Bangladesh.

Michael Sambo is a researcher at *Instituto de Estudos Sociais e Económicos*. He earned a Masters in Development Studies with a major in Social Policy for Development at the International Institute of Social Studies from Erasmus University, Netherlands, and a Bachelors degree in Economics from the *Universidade Eduardo Mondlane*, Mozambique. His research interests include transnational movements and international cooperation with a focus on South-South cooperation.

Patta Scott-Villiers is a political sociologist and convenes the Power and Popular Politics Cluster at the Institute of Development Studies at the University of Sussex, UK. She is lead author of Precarious Lives: Food Work and Care after the Global Food Crisis (Institute of Development Studies, 2016).

Alex Shankland is a Fellow of the Institute of Development Studies (IDS), University of Sussex, UK. As a member of the IDS Power and Popular Politics research cluster, he works on social movements, civil society, accountability and political representation, particularly in Brazil and Mozambique. He is also a member of the IDS Governance and Health & Nutrition research clusters and of the Centre for Rising Powers and Global Development, and has published widely on democratic decentralisation, indigenous and minority health and the role of the BRICS countries (Brazil, Russia, India, China, South Africa) in reshaping international development cooperation.

Dipa Sinha teaches Economics at the School of Liberal Studies, Ambedkar University Delhi, India. She has worked extensively on issues related to right to food, nutrition and public health both as a research and an activist. She worked with the Office of Commissioners to the Supreme Court of India on the Right to Food. She has been associated with the Right to Food campaign in India.

Lauren Sneyd is a Social Science and Humanities Research Council of Canada (SSHRC) postdoctoral fellow at the Balsillie School of International Affairs in Waterloo in Ontario, Canada. She has published articles in journals including *Food Security*, *Geoforum*, the *Journal of Contemporary African Studies* and *Sustainability*. She holds a PhD in Geography and International Development Studies from the University of Guelph in Canada.

Glossary

A Verdade	magazine (Mozambique)
Adivasi	indigenous
Bungela Mwananchi	Ordinary People's Parliament
Canal de Moçambique	magazine (Mozambique)
CanalMoz	online newspaper (Mozambique)
cesta básica	basic basket
Chapa	minibus
custo de vida	cost of living
dharna	peaceful protest
Diário de Moçambique	daily newspaper (Mozambique)
Diário de Notícias	online newspaper (Mozambique)
Diário do País	online newspaper (Mozambique)
Domingo	magazine (Mozambique)
eru	stew made with wild vine leaves, cow skin, crayfish and red palm oil
greve	strike
Indignados	literally, 'the indignant'; social movement against austerity in Spain
Ittefaq	(also *Dainik Ittefaq*) daily newspaper (Bangladesh)
Jan sunwai	social audit
Magazine Independente	magazine (Mozambique)
mbongo	dark stew made with charred wild spices and fish
Mediafax	online newspaper (Mozambique)
metical	Mozambican currency
ndole	stew of leafy greens, raw groundnuts and beef
Observatoire National	National des Droits de L'homme Human Rights Observer (Cameroon)
O País	daily newspaper (Mozambique)
Ponto Certo	online newspaper (Mozambique)
Prothom Alo	daily newspaper (Bangladesh)
revolta popular	popular upheaval
Notícias	daily newspaper (Mozambique)

Savana	magazine (Mozambique)
subsídios	subsidies
unga	maize flour
Vertical	online newspaper (Mozambique)
Zambeze	magazine (Mozambique)

Acronyms

AAM	*Adivasi Adhikar Manch*; organization for the rights of indigenous people in Madhya Pradesh, India
AAY	Antodya Anna Yojana; category of entitlement to food transfers under Indian PDS
APL	Above Poverty Line; category of entitlement to food transfers under Indian PDS
BDT	Bangladesh taka (currency)
BGMEA	Bangladesh Garments Manufacturers and Exporters Association
BIR	*Bataillon d'Intervention Rapide*; paramilitary force in Cameroon
BJP	Bharatiya Janata Party
BPL	Below Poverty Line; category of entitlement to food transfers under Indian PDS
BRICS	Brazil, Russia, India, China, South Africa
CDPM	Cameroon People's Democratic Party (English translation)
COFEK	Consumers' Federation of Kenya
CPI-M (also CPM)	Communist Party of India (Marxist)
CSO	Civil society organization
CTG	Caretaker Government
EGPP	Employment Guarantee for the Poorest Programme
EPZ	Export Processing Zone
FAO	Food and Agriculture Organization of the United Nations
FCI	Food Corporation of India
FEMEN	Feminist group started in the Ukraine
FFW	Food for Work
FRELIMO	Mozambique Liberation Front (English translation)
G8	Group of Eight advanced industrial economies
G20	Group of 20, forum for governments and central bank governors from 20 major economies
GDP	Gross domestic product
GPPP	Global Platform on Participation and Protest
IESE	Institute of Social and Economic Studies (Mozambique)

IFPRI	International Food Policy Research Institute
IMF	International Monetary Fund
LGBT	Lesbian, Gay, Bisexual, Transgender
MDM	Mozambique Democratic Movement
MENA	Middle East and North Africa
MGNREGA	Mahatma Gandhi National Rural Employment Guarantee Act (India)
NAFTA	North American Free Trade Agreement
NCPB	National Cereals and Produce Board (Kenya)
NFSA	National Food Security Act (India)
NGO	non-governmental organization
NVivo	qualitative software programme
OECD	Organisation for Economic Co-operation and Development
OMS	Open Market Sales
OPEC	Organization of Petroleum Exporting Countries
PASOK	Panhellenic Socialist Movement (Greek political party)
PDS	Public Distribution System (India)
PFDS	Public Food Distribution System (Bangladesh)
PIIGS	Portugal, Ireland, Italy, Greece, Spain
PUCL	People's Union for Civil Liberties; Indian human rights organization
RENAMO	Mozambique National Resistance (English translation)
RMG	readymade garments
RTF	Right to Food
UPA	United Progressive Alliance
USD	United States dollars (currency)
VAT	Value-added tax
WFP	World Food Programme
WTO	World Trade Organization

Acknowledgements

This book is a product of the research project entitled 'Food riots and food rights: the moral and political economy of accountability for hunger' funded by the Economic and Social Research Council/Department for International Development Joint Fund for Poverty Alleviation Research (Grant reference ES/J018317/1).

Acknowledgements are listed in individual chapters. In addition, the editors would like to thank the participants in the Food Riots and Food Rights project research, in particular the protestors and activists, politicians, public officials, scholars and experts, aid donors and journalists who gave up their time to talk to us.

We are also collectively very grateful to the research teams who contributed so much to the fieldwork and analysis in Bangladesh, India, Kenya and Mozambique: Claris Akinyi, Md. Asaduzzaman, Shreya Bhattacharya, Halima Haj, Md. Mahan Ul Haque, Md. Khobayer Hossain, Sanjida Hossain, Stephen Irungu, Benjamin N. Kiema, Eric Kitheka, Shameem Reza Khan, Mushfiqa Khanum, Cynthia Muinde, Stephen K. Mulatya, Sharlet Martha Musingi, Orah Mwende, Elizabeth Ngumii, Elisa Machel, Malombe Mulinge, Anne Mbuthia, Crescêncio Pereira, Mamun-Ur-Rashid, Omar Faruque Siddiki, Jason Waweru, and Albino Xinavane.

Versions of these arguments have been presented in workshops and meetings at the Institute of Development Studies at the University of Sussex, UK; the Critical Agrarian Studies Symposium at the Hague in the Netherlands; Center for Development and the Environment (SUM) at the University of Oslo, Norway; the Fiedrich-Ebert-Stiftung foundation office in New York City; the Carnegie Institute for Ethical Affairs; the University of Sussex in England; the University of Dhaka in India; and in events in New Delhi, Nairobi and Maputo.

Many people provided advice, support, comments on earlier drafts and participation in workshops and discussions in earlier phases of this research. We are in particular grateful to Anand, Nadine Beard, Nick Benequista, Kiran Bhatty, John Bohstedt, Hannah Corbett, Marie Jo Cortijo, Edgar Cossa, Dick Douglass, the Hon. Mohammed Elmi, Catherine Gee, Professor Jayati Ghosh, Duncan Green, Sara Hossain, Tareen Hossain, Zakir Hossain, Professor Karuti Kanyinga, akshay khanna, Iftekhar Mahmud, Dr Santhosh Mathew, Caroline Martin, Dr Santosh Mehrotra, Katy Miller, Neelabh Mishra, Sammy Musyoki, Thomas O'Neill, Cidi Otieno, Raj Patel, Prateek, Anuradha Raman, Helen Rehin, Hernán Cortés Saenz, Sachin, Nitin Sethi, Salil Shetty, Carol Smithyes, Laura Rodriguez Takeuchi, Anuradha Talwar and Emilie Wilson.

1 Introduction

Patta Scott-Villiers and Naomi Hossain

Food riots in the contemporary context

Unruly subsistence protests in low-income countries

The ten years from 2007 were marked by a tectonic shift in the tenor of relations between rulers and ruled across the world, specifically in relation to the degree to which each polity made provision for its citizens to have access to the essentials of life. In this book, we argue that the roots of the global shift in citizen-state relations lie in the politics of subsistence and struggles for natural rights, played out virtually everywhere, in different ways to be sure.

While much has been written about this shift in western democracies, a great deal less has been said about how it worked and is still working out in low- and middle-income countries. Yet, of protest events reported in international media during the period, more than half were in emerging economies (Ortiz *et al.* 2013), and it is to a selection of these that we are turning our attention in this book. Although references to protests and riots in these economies were relatively common in 2007–2008, by 2011 reporting had been eclipsed by events in richer parts of the world, in particular by the extraordinary mass occupations of public squares across western cities. When they were reported on, protests by poor people in low- and middle-income countries were generally labelled as 'food riots', in a way that assumed that their motivations, demands and implications were obvious: they came across as the uncontrolled spasms of hungry mobs demanding food. While the basic characterisation may be accurate – food (or at least subsistence) did come up as a vital component in all the events we study in this book – there is a great deal more to be understood about the political nature of the protests and how they link to economic change at multiple levels.

Given the ever-increasing degree of integration – or at least interrelation – of global economic and political systems, there is no country that is not a player in the economic changes that we are witnessing today or that was not touched by the economic crises of 2007–2008 and 2011. In 2008 in particular, at the same time as commodity prices were spiking on global markets and the financial crisis was taking hold in the high-income economies, people took to the streets to protest government failures to protect the provision of basic essentials in country after country. Their protests were suffused with the urgency of crisis and a widespread impression that politics-as-normal was excluding them as legitimate participants and ignoring their subsistence needs. Thus were the protest events sometimes uncivil, taking forms that were strident, disruptive of the social order and occasionally violent. Many met violence or uncivil responses from authorities. Some were potent, forcing official attention onto a broken moral economy: the justified belief that markets should not be allowed to eviscerate the poor. A smaller number contributed to political transformations, while others fizzled out with apparently little change to show for the risks taken. These riots and protests encapsulated a powerful popular sense of entitlement and political awareness.

The rule-breaking nature of protest events offers insight into the ways of doing politics open to people who are not part of the establishment. The defining structure of many of the protests was *political society* in the sense that Partha Chatterjee might have it, an opportunist, emotional and disorderly realm of social structuring, rather than *civil society*, its legal and more privileged relative (2004). The actions that people took often moved outside the formal modes of making claims on government and authorities to challenge the laws and policies that were failing to protect people on low incomes from hunger and to include them as persons with standing in the political system. Their protests called into question the legitimacy of ruling elites and were as often repressed or denigrated as they were appeased. In this book, we understand this mode of political agonism was part of the 'politics of provisions', a field of politics defining 'the ways in common people interacted with their rulers over subsistence, . . . permitted and shaped by pre-existing social and political networks, both among rioters and between them and their rulers' (Bohstedt 2016: 1036). We look here at episodes of citizen protest in relation to food and basic provisioning in five countries: Bangladesh, Cameroon, India, Kenya and Mozambique, between 2007 and 2012. The examples demonstrate the variation in repertoires and modes of unrest and in the moral economic arguments put by protesters and show how this variety was a function of the specificity of regimes of politics and authority in each case (Tilly and Tarrow 2015; Hendrix and Haggard 2015). They also reveal the global economic and technological forces that linked one protest to another and connected them into a global wave of contention, suggesting that the commoditisation of the essentials of life had created horizontal connections between protestors across disparate polities (most often through global economic linkages) which had, in effect, generated a single world crisis and a politics of provision that operated at local, national and international scales.

In Chapter 1, we introduce the key terms that we use throughout the book to analyse the politics of subsistence protest in low- and middle-income countries. Situating local ruptures within global economic events of the period and broader processes of economic change, this chapter argues for why food riots are a useful focus for analysis, why they have such a powerful effect on the imagination and how they play an important role at multiple levels. The chapter ends by considering the moral underpinnings of food riots and their politics, pointing to another, more profound reason for paying attention to the voices of those who suffer chronic subsistence stress and hunger.

Why are we looking at food riots?

From a world historical point of view, the decade from 2007 may prove to be a turning point in the global economy and its politics. In 2007, the extraordinary growth in financial speculation that had marked efforts to restore the fortunes of the leading capitalist agency of the epoch, the United States, precipitated the crisis that affected commodity prices and set off financial meltdown and price volatility across the world (Arrighi 2010). The political ramifications are still playing out. If Arrighi is right that the 2007–2008 financial turmoil marked the 'terminal crisis' of the US economic and political domination and the potential for a shift to another site of capital accumulation, possibly in Asia, then the unrest in many other countries can be read also as a part of this process. It may even be that many of those who protest, riot or demonstrate sense that this is a moment in which they should act if they are to stake their place in a new socio-economic regime. Whether this is reading too much into it, the commodity and financial crises certainly did unleash hitherto quiescent stresses among masses of informal and low-income workers and peasants in a wave of unrest across the world. While, in part, an expression of grievance at immediate conditions, we believe that they were also attempts to assert a political subjectivity that had long been denied.

The protracted crisis of subsistence brought millions of people out onto the streets. The Global Platform on Participation and Protest[1] records protests against corruption, unaccountability and failures of representation in 2007–2008 in Bangladesh, Bolivia, Brazil, Cameroon, Kenya, Madagascar, Nigeria, Pakistan, the Philippines, Slovenia and Tunisia; and food related protests in Argentina, Bangladesh, Burkina Faso, Cameroon, Egypt, Ethiopia, Haiti, Côte d'Ivoire, Kenya, Mauritania, Morocco, Mozambique, Nicaragua, Nigeria, Peru, Senegal, Somalia and Uzbekistan. The number of protests then increased year on year (Ortiz *et al.* 2013). Chapter 2 of this volume, *A world in protest* by Sara Burke, draws on the Global Platform to describe the evolution of protests from 2007–2012, observing an evolving pattern as different political economic phases and geographical manifestations succeeded one another.

A key task is to clarify meanings. What assumptions are we making when we talk of 'food riots' in the early 21st century? What judgements are applied to labelling an event as a 'food riot'? What are the political effects of this labelling? Sara Burke's chapter suggests that protests came in many forms and types during this period, usually combining material and political demands. There was a geographical divide: we have already noted that what were called 'food riots' were usually recorded in low- and middle-income countries, the latter often under authoritarian regimes. Yet the characteristics of some protests against austerity in Europe, such as the 2011 riots in the United Kingdom for instance, closely resembled struggles in poorer countries over food prices. And wage struggles in the south were more like the industrial disputes of the north than they resembled a struggle over the moral economies of grain in the south. Many food-related collective actions in the south tended to be neither violent nor particularly transgressive, more protest than riot and often involving participation of people close to power and advocating policy reforms. The Right to Food social movement in India is one such example.

So what then do we mean by 'food riots'? We mean the unruly actions of the crowd to assert to 'authorities' a right to food. We mean political actions that counter the usual order and whose participants legitimate their claims with reference to a moral economy; specifically, a shared understanding that food markets should work for people. It is these individual moments, spikes of civil transgression, on which we centre our analysis in this book, because they function, in the systems by which poor people do politics, as signals of distress and of a need for recognition in the formal register of politics. These units of analysis build up into episodes and processes, many of which combine into wider struggles over the responsibilities of state and the rights of markets and people. These episodes, when they affect the food system, comprise the politics of provisions. So we talk a great deal about food riots in order to make sense of them in light of the negotiations, interests and repertoires at stake in bringing about and maintaining relations between rulers and ruled that are functional for human wellbeing and social reproduction. The media play an increasing role in shaping this relationship and shaping our ability to understand what is happening, a subject to which we will now turn, and which we delve into in Chapter 3.

Why media framing matters

Much of what we understand about food riots is drawn from the framing provided to us by the media. The 'spasmodic' view, though roundly criticised by Thompson (1971), Bohstedt (2010) and C. Tilly (1975), resonates widely in the reportage of the international media; this may be a function of the brevity with which the story 'hungry-angry-crowd' can be told through the term 'food riots'. This term has an impressive English-language communicative capacity to telescope complex politics into a newsworthy event. Initial analysis suggests this is a key framing used by the international media.[2]

The term 'food riot' in the international media carries a powerful external accusation, based on a reading of the conditions and a rough knowledge of history, that enables an assumption that 'food riots' signal a precarious regime, economically and (therefore) politically. It highlights to an international audience in geopolitical space the apparently fragile legitimacy of a regime in which the events have occurred. The name also sends another, more powerful signal to the international political economy. As Chapter 3, *Media framing* by Hossain *et al.*, suggests, the siren call of global economic and financial agencies is that freedom works; they seek to mobilize support for freer trade and agricultural industrialization as the surest route to food security, with emergency food aid to fill gaps. Markets need to be enabled to provide. In stories of riots such agencies may also read a threat to the long-term viability of their economic models, recalling perhaps the 'IMF riots' of the 1980s in Latin America and elsewhere (Walton and Seddon 2008). This perception of a threat is accurate enough: protestors protest about the results of neoliberal food economics, even if they do not always point a finger towards the specific global or regional agency.

The labelling of some, but not other, events as food riots also distinguishes between the complex social and political and economic realities of riots or protest struggles in the north and the (implicitly simpler) rumbles of the belly in the south. This is a largely false distinction between these waves of economic protest that northern activists also often ignore. There is an interesting difference, for instance, between the moderately sympathetic view of the international media towards food riots in southern countries uncovered by the review in Chapter 3 and the generalized media hostility towards the London youth riots of summer 2011 (Cooper 2012; Bassel 2012). But in the reported looting of trainers and tech goods, young people were protesting their inability to possess the things that they must have, as citizen-consumers, in order to be a member of 21st century society. Perhaps there is no moral equivalence between the destruction or looting of private property for consumerist 'loot' rather than in order to eat. But what this distinction between necessities for a socially-determined good life effectively does is to certify one kind of protest (around food, among poor people in the south) as morally necessary, and justified by the corruption of those regimes. By contrast, broadly parallel events, by people in similarly constrained and unequal circumstances in London, were scorned as the result of crass consumerism at its most violent. The London rioters were thus de-certified as legitimate actors in a struggle.

It is important not to discount the importance of food vs. trainers as a moral matter in these events, but also to recognize that if there is a moral difference between the two cases, it is not very important in practice. We know this because even if some food rioters were actually hungry, many were managing food one way or another, and it was not usually the hungriest groups or even those hardest hit that protested. So while food does matter, and it may make a moral difference to the politics at play, we have to steer clear of assuming that is because the belly rumbles. Food matters here, but in a political and not a physiological way. Other basic needs of everyday life may matter equally, and in the same ways.

Professional reportage (rather than social media) tends to take a strongly moral stance, which reflects and shapes the political mood. This influence is not confined to the broadcast or press media, but also applies to scholarship, whose interpretations of events always contain a degree of normative rightness and wrongness, admiration and denigration, however nicely worded. Understanding what threats, entitlements, and meanings are involved are vital in shaping response to protest claims, complaints and challenges. This is particularly important in relation to notions of security to which we will now turn.

Food and other securities – the riot in imaginary

In development policy research, food riots are causally and consequentially associated in the relationship between food security, conflict and human security. In the surge of studies of food riots that followed 2008, many situated explanations within debates about violent conflict, studying associations between protests, demographics and food prices across economic and political settings (Demarest 2015). Brinkman and Hendrix include food riots among the list of forms of (violent) conflict associated with food insecurity, the full list including civil conflict, interstate war, democratic and authoritarian breakdowns, protesting and rioting, and communal violence (2011).

O'Brien notes that the 'food riot is an important marker of insecurity and discontent', blending 'different forms of insecurity, combined with a desire to seek redress for perceived injustices' (2012:44–45). His analysis views rioting within the wider continuum of political struggle and its relationship to human and food security. The continuum suggests that food riots may unleash further unruliness, particularly if they elicit state repression, but also if they raise and dash mass expectations. As Hendrix and Brinkman note, even positive responses to food riots create 'expectations that food price interventions will continue, and encourage consumers to evaluate the government explicitly in terms of their ability to stabilize consumer prices' (2013: 9). However, this is not necessarily always the case, as the five examples in this book will show.

Specific episodes of struggle such as the Egyptian 'bread riots' of the price spikes era have been linked to 'threshold' price rises beyond which violent conflict, or mass political struggle, is believed to be more likely (Lagi, Bertrand and Bar-Yam 2011; Bar-Yam, Lagi and Bar-Yam 2015). The 'threshold price' argument resonates with Bellemare's conclusion that price rises create the conditions for protest (2015). Although she notes cautiously that 'interesting associations between food prices and social unrest . . . are just that, and in no way do they imply that food prices actually cause social unrest', Bellemare goes on to conclude that high – but not volatile – prices probably do lead to social unrest (2015: 12). Demarest observes that the conflict-food riots literature embeds 'an implicit assumption that the causal chain between increasing international food prices and conflict runs through food insecurity'. With a reference to the 'spasmodic' view of food riots, Demarest critiques correlational analyses of insecurity and food, noting that these are embedded in an 'image of a rise of the hungry poor' (2015: 15). As she notes, broad correlations between conflict and food security give us some associations with wider social, economic and political processes associated with food riots, but get us little closer to any mechanisms of cause and effect that might justify an interpretation of food riots as leading directly to problems of national security. Bush and Martiniello argue that:

> The conjunctural reason for conflict may be a spike in food prices but it is a mistake to see famines and food insecurity more generally without the important historical framing that highlights the structural and longer-term reasons behind persistent protest.
>
> (2017: 104)

Localized struggles can diffuse into wider conflicts across a number of mechanisms, including through commercial and popular media, which may disseminate accounts from one discontented population to another, or it can be imagined having such an effect. This gives reasons to governments to suppress unrest, through appeasement or repression, and will also give a degree of priority to broad food security measures. Riots are always dangerous, as E.P. Thompson (1971: 120–122) pointed out, involving risk to individuals and often decades of problems for communities. People and authorities usually try to avoid them.

The politics of provisions

National subsistence settlements

This section of the introductory chapter unpicks some of the essential workings of the politics of provisions and ties them into broad processes within the global economy. We can think of the politics of provisions as arriving at 'subsistence settlements', borrowing from understandings of how more or less inclusive economic growth may be enabled by 'political settlements' (Hickey 2013). Powerful contending groups agree in some manner on the terms and manner of the distribution of power and resources; in this case, subsistence resources. The overall balance of power is shaped by the 'holding power' of different groups in the ever-contested equilibrium, which may include mediating demands from constituents who are less powerful individually, but together and with leadership constitute sources of power. Whether the overall settlement is a competitive one or is dominated by a single party will affect who it includes, and how it enhances or constrains change and in which domains. The extent to which relations between rulers and ruled are clientelistic or impersonalized also shapes these outcomes, affecting which policies are chosen and how they are implemented. In addition, the extent to which the substantial powers of the market are unleashed or controlled within the settlement affects its stability; for example, the degree to which economic regulation or other protectionist activities are included.

We have suggested that states have a general interest in ensuring the population is sufficiently well fed to reproduce itself. There have been populations for whom this was not true, including rebel populations, and that has been the story behind a number of famines in the recent past (Biafra 1970, Ethiopia 1973–1974, among others [See, in particular, De Waal 1997]). But the general point is that states prioritize food security, and will make arrangements to protect it if they can because politics are built on these foundations. Not only do subsistence crises have a wide range of undesirable political economic impacts, from the point of view of elites – they may lead to population displacement as people look for food or work elsewhere, depopulating political constituencies and raising the cost of agricultural labour — but they also often lead to civil unrest, including violence, particularly in urban sites where affected populations are in close community and are quick to feel the effect of 'triggering' events such as policy announcements or police crackdowns. Urban unrest is associated with politicization among affected groups and shifts in elite power bases. For essential populations to suffer would also signal a state unable to honour its most basic functions, inviting geopolitical interventionists, while food riots can give the appearance of regime vulnerability, an invitation to contending elites.

Charles Tilly has shown how the modes by which western elites historically sought to manage subsistence challenges changed with the growing complexity of the state and society, including the expansion and extension of markets and the state (1975). Army provisioning was a major drain on the food system, but in wartime doubly so. Bureaucracies, the middle classes and, as de-agrarianisation proceeded, the urban masses all needed to be fed, regularly and well. As economies developed, state priorities tended to move towards greater food industrialisation and market integration. Agricultural productivity rises and market integration meant that progressively, the food system was able to improve people's access, and the old forms of food riot died away with the peasantry. With more urbanisation and food trade, blockages and price riots came to be more common forms of food contention, reflecting the more commodified relationship between people and food, and (therefore) how the state-society relationship had changed.

The marketing of grain, changing with agricultural production methods and demand, is a node of conflict throughout European histories of food policy. Louise Tilly has shown how conflict in France was linked to specific changes in the marketing of grain (1971). E.P. Thompson's 'English crowds' similarly contested changes in the rules of markets for staples, or the failure by the public authorities to abide by them (1971). The present-day association of contemporary food riots with global price spikes speaks similarly to this theme of a politics of provisions, which is a politics of how food markets should work.

Charles Tilly periodizes the varieties of food contention in Europe, but others find it less easy to identify a sequence of food protest repertoires in the global system, rather finding them more plural in nature, reflecting the diversity of relationships to food markets, of political modalities and contingencies, and unevenness in the degree of integration into the global food system (Taylor 1996). The sample reviewed here in Chapters 4 (Bangladesh), 5 (Cameroon), 6 (India), 7 (Kenya) and 8 (Mozambique) reveal no clear pattern of protest repertoire: there are price riots, wage strikes, blockages and 'classic' food riots (taking over the market itself), as well as more polite forms of civic action and media activism identified in this recent sample. The politics of provision is affected by capabilities, notably horizontal relationships between those who are suffering food shortage, and vertical relationships between the people and authorities (Bohstedt 2016). The Chapter 6 case study on India by Joshi, Patnaik and Sinha shows an example of a vertical relationship in transition, when the waning powers of the Communist Party (Marxist) (CPM) that had ruled the state of West Bengal for decades was the political moment for a series of intense protests by farmers and people on low incomes around the management of the food subsidy system, a structure infused with CPM politics across the state. This case is contrasted with an example of a remarkable 'horizontal' relationship, whereby Adivasi tribal groups in Madhya Pradesh, threatened with rising levels of malnutrition, protested in alliance with community organisations, larger civil society groups and lawyers to generate state responsiveness to their exclusion from welfare and essential services. The story of the Right to Food movement in India shows the politics of provisions encompassing ruly and unruly approaches, arriving at and sustaining an evolving subsistence settlement.

The kind of responses to crisis that indicate that politics of provision were at work were evident on a national scale in all our cases, from repression in Mozambique (Chapter 7) to rapid response in Bangladesh (Chapter 4). Equivalent mechanisms of response were also visible at a global scale, for instance, in the way the global food aid system was substantially mobilised in 2008 (Bohstedt 2016; Clapp 2012; Lentz and Barrett 2013). In addition, individual countries took specific actions in the global economy – as we show, for example, in the cutting of import tariffs by Cameroon in Chapter 5 and Kenya in Chapter 7; while India, as we note in Chapter 6, cut rice exports to protect internal supplies. These kinds of action demonstrate the degree to which a given politics of provisions has achieved an automatic right to protection for a country's entire population. Understanding how this has come about in each case involves historical analysis, including contribution or acquiescence to economic trends over the long term. In the following sections, we will look at some of the major driving forces in the global economy that have been the result of decades – if not centuries – of back-and-forth political bargaining within and between states and market actors and which are arguably key to even the most local issue of provisioning.

Global food regimes

Food regime theory offers the most systematic explanation of the struggles generated by contradictions in the corporate food regime, and of the implications of global food price spikes, inflation and price volatility for different populations. Food regime analysis explains the evolution of the world's food system in relation to the needs of capital expansion, identifying the associations of capital to 'particular configurations of geopolitical power' (McMichael 2009b, 2009a, Friedmann 1993). In the historical analysis required by food regime analysis, there are distinctively different arrangements for the production, finance, distribution and use of foodstuffs over different parts of the world, depending on the needs of labour, war or security and technological development. The food system, the foundation of social reproduction, was structured around core economic functions like the industrialization of cereal production, the green revolution and agricultural intensification, commoditisation of food crops, and the provision of cheap goods for the low-waged (McMichael 2009b). The first global food regime was colonial, with food production outsourced to colonies to feed growing European industries. The second regime saw the growth of transnational agribusiness between national farm sectors based on the national model of economic development. The third is the corporate food regime, in which mass production, standardised foods and global supply chains dominate. Philip McMichael claims with considerable justification that each regime and its transitions has reshaped the politics of development as well as influencing the direction of technological and social change. He argues that when, for instance, millions of small farmers are forced to leave the land as it no longer pays in competition with global agri-business, they become a 'reserve army of labour' for such new complexes as export processing zones, and thus the food regime illuminates not only the history of capitalist food relations, but also the history of capitalism itself (McMichael 2009a: 281).

Predictably, the precise nomenclature and periodisation of regimes is under continuous development, but the idea remains useful to us nonetheless. A post-2008 food regime has been characterized by an 'end to cheap food' (Moore 2014; Patel 2012). What has this meant in practice? High and rising food prices worsened the imbalance between work and earnings among the already low-waged. Rising fuel prices also increased the costs of farming and provided a source of particular grievance to people on low incomes who had to travel for their livelihoods (Hossain, King and Kelbert 2013; Hossain *et al.* 2015). Volatility and low wages are built into the corporate food regime, evidence of the 'freedom' of global markets at work (McMichael 2009a). This fundamental contradiction, whereby 'free markets' exclude and/or starve populations dispossessed by their very implementation, characterizes the corporate food regime and is one of the sources of today's food riots. The argument is that the era of globalization has been premised on the food regime's generation of cheap labour for manufacturing, service and industrial agricultural sectors, and its supply of relatively cheap industrial foods to subsidize labour costs (Patel 2012).

Global food price crisis

In mid-2008, the world wholesale price of a tonne of wheat stood at USD $140, triple its price of the previous year. Food prices were 83 per cent higher in 2008 than they had been in 2005 (Bush and Martiniello 2017). The dramatic rise of food prices on global markets that began in late 2007 and peaked in mid-2008 was followed by an equally dramatic plunge. In 2010–2011, commodity prices spiked again. The causes of global food shortages were attributed to poor harvests in the main producing areas of Ukraine, Russia and the United States, high fuel prices that affected fertiliser use, and diversion of cropland to biofuels (Holt-Giménez 2008). The price volatility that turned these shortages into a crisis of food availability appears to have been caused by the enmeshment of global agricultural systems with the global financial system (Ghosh 2010).

It has been plausibly argued that the reason for the sharpness of the 2008 spike was not only because of production problems in key producer regions, but also because, with deregulation of commodity markets in the early 21st century, capital was able to flood out of unstable subprime housing markets and into commodity futures and spot pricing as investors looked for new markets to which derivatives could be applied (Holt-Giménez 2008). The volume of this move raised prices of food and energy with startling rapidity, and when the money was shifted out of commodities with the banking collapse, prices fell with equivalent force.

It was the concatenation of food, fuel and financial turbulence that transmitted to retail prices in the developing world. Prices shot up almost everywhere, even in countries that depended only to a small degree on food imports, and, unlike prices on global markets, retail prices then tended to stay high (Dawe and Morales-Opazo 2009; Ghosh 2010; Scott-Villiers *et al.* 2016). While the global food and financial crises seemed to be over by 2012, at national level the crisis continued in a new guise of chronic subsistence stress and sustained political unrest. This was the case in the global south and north alike, manifesting differently in each political economic context. The spikes had indicated not merely an immediate price problem that could be managed by protection measures (regulating banks, protecting industries and people), but also a deep structural crisis in which mechanisms of social reproduction were fractured (McMichael 2010). The protests that followed indicated veritable fear that the political economy of provisioning had come adrift from natural justice (Patel 2012).

In the period leading up to the crises of 2007–2008, the world economy had become increasingly tightly integrated. Trade between industrial economies and emerging markets increased dramatically (IMF 2007) and international remittances from rising numbers of migrant workers criss-crossed the planet.[3] When, in 2007, commodity prices began to peak worldwide and labour remittances fell, the effects were transmitted from one country to another. The economic links emerged in the country contexts elaborated in this book in different ways. For instance, Bangladesh, became the second largest exporter of cheap ready-made garments in the world in 2011 (Ahmed and Nathan 2014). In 1998 the industry represented 15 per cent of the country's exports, but by 2011 garments comprised 50 per cent of exports (*ibid.*) and by 2013–2014 it had reached 81 per cent.[4] The contracting world economy exerted downward pressure on wages and conditions, while worker struggles and western consumer concerns created a counter pressure. In Africa, meanwhile, Mozambique, Kenya and Cameroon posted increasing levels of international investment in energy and agriculture and, with continued dependence on export of basic commodities, felt the effects as prices rose and fell (UNCTAD 2015). In particular, the volatile prices of food had deleterious effects: even though larger producers and traders benefited when prices shot up, the majority of the populations in developing countries suffered a dramatic loss of purchasing power.

It appears ironic that as the global crisis worsened, aspects of market integration deepened among the less strong players. Local industries had to make price concessions in their struggles to survive wildly volatile prices, falling demand and increased competition. Increased integration with global markets at a national level was mirrored in the reaction of millions of people on low incomes who, faced with rising prices of essentials, went out to raise cash in any way they could, through taking on every kind of work regardless of conditions (Scott-Villiers *et al.* 2016). With little time, space and energy, they also turned to the market for the cheapest food. Their consumption and labour became, in effect, commoditised through the effect of the crisis on their bodies and lives, in a process by which 'the struggle to contain the consequences of commodification [led] to the further extension of the market' (Burawoy 2015: 24).

Labour value

As analysed by Farshad Araghi (Araghi 2009: 114), the crisis of global value relations is expressed in new time disciplines and demands and lower wages: the very value of labour is at stake. How do people experience these moments in the process of 'accumulation by dispossession' that dramatically alter the balance between work and the basis of life? The present authors have been exploring these issues in related research in ten countries, some of it spanning the years 2009–2015. This research tried to gather interpretations and accounts of the changes in people's lives after the period of spiky and volatile food prices. This work covered three years after the last major global price spike, and listened closely to the views of people about how they had adjusted to the realities of considerably higher real prices. Some of that research conveyed a very strong sense of squeeze: a combination of high food costs across all categories, other demands (school expenditures, rent), and stagnating or insufficient rises in wages (Hossain, King and Kelbert 2013). Although in 2017 the popular complaint against inflation has dissipated, it has not disappeared but has been channelled into complaints about labour value and opportunity (Scott-Villiers *et al.* 2016). Life has substantially adjusted across broad swathes of very diverse populations, and people do more and different types of work than in the past, often travelling and juggling considerably to keep it all going.

During the years of crisis, low income wages did not automatically adjust to the higher prices; rather, it took time and collective action for change to come about – and it is arguable that wages never caught up with the rising cost of living for the poorest. Much depended on how people were differentially affected by macroeconomic forces. Groups like public administration staff, with generally strong organizational and strategic presence, were in most countries relatively better protected against inflation than those in the more precarious parts of the private sector. The evidence on how people viewed inflation suggests a mixed story: that poorer people in general disliked inflation but the extent depended on the relationship between inflation and unemployment, and how it affected their country and their individual efforts. Easterly and Fischer argue that inflation more often tops the list of major national problems for poorer than for richer people, and that there are good reasons for this: poorer groups in general suffer more from its effects (Easterly and Fischer 2001). Evidence from twenty advanced countries took a cyclical view of popular macroeconomic priorities: rising and volatile inflation was 'more costly, and the public put greater emphasis on low inflation as prices increase more rapidly', with great national variation, depending on the aggregate costs of these macroeconomic policies under different governments. But people similarly became more concerned with unemployment when that was on the rise (Scheve 2004: 2). Scheve concludes that '[t]he role of distributive conflict among groups in society is central to debate about this question [of inflation]' (*ibid.*:29). In our terms, how much inflation a society would tolerate was one of the macroeconomic issues negotiated as part of the politics of provisions.

Still other evidence concludes that poorer groups dislike unemployment relatively more than they do inflation, unlike the rich (Jayadev 2008). Jayadev argues that this can be explained by class conflict theories of inflation aversion, reflecting the class interests of the rich and the poor. But time horizons also appear to matter, because how people judge the present relates to what they expect to happen in the future, based in turn on their knowledge of what has happened in the past. Analysis of Gallup World Poll data leads Gandelman and Hernandez-Murillo to conclude that inflation and unemployment

> depress the evaluation of the present relative to the evaluation of the future. In other words, higher current inflation or unemployment creates the effect of improved optimism, not because the future is assessed as more favorable but because individuals believe the present looks grimmer.
>
> (2009:119)

It was understandable that high levels of unemployment and inflation, as these went together during the global food crisis, should create a sense that the present was particularly grim. But the implication of Gandelman and Hernandez-Murillo's findings is that people would view unemployment and inflation as necessarily temporary situations. Perhaps this reflects the intolerable nature of a life living squeezed on low pay and ever-higher costs, the impression that society cannot go on indefinitely in such a way. For individuals interviewed in the *Life in a Time of Food Volatility* project[5], the sense of a squeeze included important elements of economic psychology: that higher real expenditure made it impossible to save or to plan; a sense that wage rises were illusory, as wages bought less than in the past; and a sense of not getting better off, despite working harder. Very few people could say their earnings were keeping pace with price rises, even as late as 2012 (Hossain, King and Kelbert 2013). People who had been involved in protests and riots explained that they rioted not because they felt at a dead end, but because they needed to speed up the pace of their incorporation into a rights-bearing position in society.

Where wages and earnings proved 'sticky', some groups had to struggle to bid up wages. Rickshaw pullers in Bangladesh, for instance, reported daily arguments over fares as prices rose and passengers sought to keep transport costs low; at one time during the 2011 price spike, they went 'on the rampage' when their rights to ply certain main roads were removed (Rashid, Hasan and Hossain 2012). Chapter 4 shows how garments workers' protests in Dhaka were closely linked into the problem of inflation, which had seen their real wages shrink during the 2000s. There were similar expectations of having 'to come down to the street' to demand lower food prices, suggesting in addition, an expectation that to do so might be effective, if feasible.

It is a tenet of economic theory that 'wages adjust' with inflation, and so – hopefully before long – everything turns out all right for consumers, but the invisible hand of the market is not always very quick or very fair in its distribution: wages take time to adjust; how they do adjust – whether through statutory inflation-indexing pay or through piecemeal bargaining on the road or the plant floor or through riots – may reflect their political or economic importance more than how workers' needs may be changing as a result of higher real costs of living. And of course, there is a growing excess: the neediest and most informal may not be sufficiently empowered to bid their wages up, and may have already been dispossessed of the land, assets and education they might use to achieve their aims. Add to this the possibility that it has become harder for the billions of people at the base of the pyramid of global inequality and their governments to achieve any leverage on a highly financialised economic system (Standing 2016) and it appears that perhaps wages do not really adjust for everyone. For one thing, the costs of wage adjustments are recouped by employers in the colonisation of the workers' unpaid time: work today can involve more time, less freedom, harder labour, more risky migration, more waiting around and higher physical and moral risk (*ibid.*). The costs of such adjustments are real and enduring, and may be adverse for long-term wellbeing. For instance, as Sneyd explains in Chapter 5 and we also learned through our work on the *Life in a Time of Food Price Volatility* project (Hossain *et al.* 2015; Scott-Villiers *et al.* 2016), some of these adjustments indicate moves towards reliance on increasingly industrialized, low-nutrition foods – suggesting a potential hazard, now or soon.

The double movement: protection of the social

Struggles over the key elements of the subsistence settlement – food regime, labour value, inflation, unemployment – and political conditions, histories of contention and structures that have built up as a result help us to make sense of food riots, and how and under what conditions they 'work' (Tilly 1975; Bohstedt 2010). We think it is only possible to see the change such events bring to the marketing of food and to state arrangements when we consider them within long economic cycles which themselves contain cycles in the relations between protestors and public authorities. McAdam, Tarrow and Tilly (2001) make a persuasive case for exploring longer and wider periods of contention through study of how individual events shift, confirm or widen the focus of contention, connecting protestors of similar profiles with each successive articulation of grievance. Chapter 9 uses this insight and looks back over the evidence presented in the book to uncover the mechanisms that work to bring about such change, and how these sequence and interact with one another and with processes of social and economic change (such as urbanisation and de-agrarianisation). These processes and events are embedded in varying combinations into long haul struggles over the distribution of power and resources.

Are we witnessing, in this wave of protests, a resurgence of Karl Polanyi's double movement in response to the dangers of economic liberalism (Block and Somers 2014)? Polanyi argued in 1944 that a nation that hoped to govern by free markets was a 'stark utopia', which 'could not exist for any length of time without annihilating the human and natural substance of society; it would have physically destroyed man and transformed his natural surroundings into a wilderness' (Polanyi 1957: 3). In the light of this, there would be either backlash or adjustment to protect this substance. Since liberal markets could not by themselves provide the necessities of social reproduction such as good food, air, water, education, health and other care, fair treatment in work, and social and personal security, this would be done through politics. Money, labour and nature, said Polanyi, are false commodities, not things that can be ceaselessly exploited with the ebb and flow of demand. When these elements of the lifeworld are subjected to market principles, he proposed, social relations are threatened and major crises to economy and environment will ensue. Polanyi believed that a movement towards protection would result, originating in suffering, urged on by disaster and riot, and occasioned by the fears (or foresight) of industrial leaders. States would move to institutionalise redistribution, recognition and representation (Fraser 1997), in part because the suffering multitudes would force attention to their needs.

Some have questioned Polanyi's optimism that a double movement is automatic. As each phase of capital integration succeeds the last at a higher scale of complexity and global integration, it becomes less obvious how a single state or group of states could evade the logic of the rapacious market and enforce regulation and social protection. Burawoy, for instance, suggests that while Polanyi was right to point out the inherent instabilities of laissez-faire and impulses to counter movement, the likelihood of state intervention in setting matters to *full* rights never was a given (2015). Polanyi thought that states acted on ideas. Burawoy argues that they act on the needs of the moment. He points out that today, 'a countermovement would have to assume global proportions, even as it is built on local and national grievances'. States, as market players, are embroiled in keeping the markets happy, and ordinary people are also enmeshed in markets for their survival. This requires of us an inquiry into what actually might make the double movement work in today's circumstances, a question to which we will turn in Chapter 9, the concluding chapter of the book.

The moral economy

A key puzzle is: why were people in all sorts of places not known for mass protest suddenly in the streets when prices spiked, often despite the likelihood of state repression? What is it about food that might make the double movement begin to function again to protect the people from the ravages of a feral economy? As a first step, we may turn to the classic work of E.P. Thompson on the moral economy of the crowd. His exploration of food protests in 18th century England demonstrates that 'the men and women in the crowd were informed by the belief that they were defending traditional rights or customs; and, in general, that they were supported by the wider consensus of the community' (1971: 78). Food riots then were triggered by soaring prices and real or perceived malpractices by dealers or millers. The impetus to break rules was grounded in the crowd's view of social obligation between rich and poor in that society, an obligation that was largely shared by the nobility, if not by the industrialists who were benefiting from the flood of labour into their newly built factories and the plutocrats and dealers who were engaged in international trade in corn and causing prices to rise at home. Thompson's research shows that the behaviour of the crowd tended to have clear objectives to draw attention to rights and customs, and acceptable degrees of unruliness; in other words, it had its own morality. It was deemed to be part of a greater morality relating to fair distribution of the essentials for life. The notion of a moral economy that informed protesters and rioters shows up in our case studies as applying as much today as it did then, even as the notions of what is fair; what part must be played by citizen, market and state; and what tradition of governance and law is being called upon has changed and varied with each successive era of capitalism and in each of its conquered territories. It is a marked feature of moral economic protests that there is often a call for reinstatement of an old way of controlling the rapacity of the market, not recognising that capital has already moved on and a new phase of relations is now in formation. Nonetheless, it would seem that the moral economy serves as the engine for the politics of provisions, in which the crowd enunciates both its power and its sense of rightness, to authorities that are, in some senses at least, 'the prisoners of the people' (*ibid*.:79).

If we are to understand why the moral economy persists through repeated cycles of change and across so many different polities and may have the power to call forth the double movement at a global scale, we need to go deeper. Beneath the straightforward rightness of a morally regulated economy lies a deeper moral issue relating to the value of life and its future reproduction. Morality in the end is always about what is good for all, not only some. Contemporary protestors, north and south, are aware that nature, of which people present and future are a part, cannot be infinitely exploited for the benefit of some small part of itself, a few well-endowed human beings and their capital arrangements (Moore 2015). The moral idea of 'good for all' justifies a primordial righteousness when this natural law is violated to excess. In the industrialised countries, this was revealed in opposition to the primacy of financial capital over social and environmental life. In emerging economies, it showed itself in noisy protests calling for fair treatment within a fairly modulated economy.

Damage to people and environment has been a feature of the decline in each successive capitalist regime, as Polanyi (1957) pointed out. When the exhaustion of the worker and her lifeworld reaches extraordinary proportions, something has to give. This seems to be a description of the contemporary moment, the time of signal crisis of the neoliberal age (Arrighi 2010). Exhaustion is the declining ability to squeeze more productivity out of the industrial workforce, the dwindling of new sources of unabsorbed labour and the worn-out worker herself, who provides labour to multiple jobs as well as contributing impossible amounts of unpaid effort to caring for family, home and community, and to administering the conditions of her precarious job arrangements (Standing 2011; Arrighi 2010: 224).

The moral power of critical events

The signal crisis is a critical event, in the sense that it has generated mass suffering and will institute new modalities of political action (Das 1995). In country after country, the stresses on access to food and employment have changed the geography of ordinary life. What was once local for many of the people met during the research undertaken for this book – the neighbourhood, the local shops, the farm, the town, the city – was suddenly juxtaposed with much more distant markets forcing prices up and wages down, leaving people hungry and sweeping away traditional social protections. National institutions were forced into the arms of global institutions in more or less intricate ways: India did battle with the WTO to be entitled to subsidise its agriculture as the United States and Europe are allowed to do by its arcane rules; Bangladesh was heavily subsidising its ready-made garment export industry to keep it internationally competitive; Cameroon, with international help, flooded its urban food markets with subsidised Thai rice; Kenyan leaders tried to avoid the indictments of the international criminal court for post-election violence fuelled, at least in part, by international land grabbing; and Mozambique came to tacit agreement with international donors to subsidise bread and fuel and with international companies to lease out large tracts of valuable agricultural land.

New identities emerged as people moved to protest loss of once-secure assets, dwindling patronage and erosion of traditional identity: in rural West Bengal, farmers and labourers contested their bureaucratic categorisation and insisted on new political identities. Meanwhile, in Madhya Pradesh, a civil society movement to implement a constitutional right to life and its corollary right to food was joined by malnourished tribal castes whose identities were transformed from isolated and dispossessed categories to deserving citizens of a state that connected rural villages to the Supreme Court. In Cameroon and Mozambique, rioters were treated as enemies of the state, consolidating their growing identity as an embattled urban class.

Veena Das argues that suffering acts as a narrative trope during such critical times, by which the state appropriates the suffering and claims the moral ground (*ibid.*:18). It labels, categorises and eventually, in some way, blames the victims. They may be assisted, relieved, reviled, punished or promised (and we give examples of all of these in this book), but the truth of their experience is attenuated. During the global food crisis, millions of people on low and precarious incomes experienced and continue to experience hunger and a daily struggle to maintain families, houses, farms and employment. Their experience has been abstracted into global reports and nutrition figures, their status as human beings and political citizens reduced, and their voices silenced. Their poor performance in the economy has been ascribed to a failure of will and skill, rather than a massive structural disadvantage.

Thus comes the logic of the riot: not beseeching, nor complaining, but a demonstration of presence of 'the inexistent' and a reminder of the truth (Badiou 2012). Riots carried the pain of the common people and the cry for protection into the public space. They also brought, to be sure, the internal power dynamics and violence that flocculate around leadership and voice as much among the dispossessed as among any other surging group of humanity. Issues of moral right, repertoires of unruly behaviour and episodes of internal and external coercion and violence played out in the riots and protests. The 'community' (those who were suffering) became a new political actor influencing the acts of individuals within it (those who protested and those who did not), and in contention with other political actors, including those in the market, affecting prices and access, as happened in Cameroon; in the courts, by raising claims regarding the right to life, as in India's and Kenya's public interest cases; and in the state itself, when activists found political opportunities to push for changes to constitutions, laws, policies, programmes and political discourse.

Das calls for public rituals that would allow the reality of suffering to be properly incorporated into politics, a move that could make the double movement more likely to come about and the moral economy moral once again. She does not believe in the 'simulation' provided by newspaper articles, government reports, legal judgements or statistical categories, but instead calls for *incarnation*. She gives an example. In 1990, six years after the Bhopal poison gas tragedy, the incoming Janata government, bypassing (but not contravening) the labyrinthine legal and bureaucratic structures that had been set up to categorise victims (and ended up blaming them), establish their entitlements and make claims on Union Carbide (the company responsible for the fatal leak of methylisocyanate), made available a stipend to each of the 106,000 households affected for a period of three years. Das explains:

> [A]t a stroke the victims were transformed from incompetent, irresponsible people to those whose needs for survival were of paramount importance. Above all, the limits of judicial and bureaucratic rationality were recognized by the state and it decided to absorb the uncertainties itself rather than make victims bear the costs. . . . It was a rare instance of a politically humane moment that needs to be recovered, for it affirms existence by its very 'irresponsibility.'
>
> (Das 1995:172–175)

In the chapters that follow this one, we will hear how people called for this political humanity in waves of protest. We begin by looking at protests across the world during the period in question in order to understand the larger scope of the signal crisis of which local riots are a part. Then we turn to a consideration of media framing in order to demonstrate not only the considerable power of media in the politics of provision, but also the tendency to miss the point about what truth the riots offer – a tendency that we hope to overcome in this book. Our five country cases show instances when the ruptures prompted instances of political humanity and demonstrated the attempts to give birth to a new moral economy. They also show how moves towards recognition and redistribution were counteracted by other measures that favoured the free market, in part because of vested interests and political calculus, but also because most sections of society in our five case countries depended on the market.

Notes

1 See the "Global Platform on Participation and Protest" (GPPP) http://global-platform.org, an evolving global documentation of contentious politics between 2006–2013 based on news and analysis openly available on the internet and still a work in progress.
2 Dangers of telescoping notwithstanding, news sources can be very useful, as Burke's study shows in Chapter 2. The database created by Burke and her colleagues involved careful hand searches to build up an account of particular events within their wider struggles, keeping in mind the advice of Tilly, Tarrow and McAdam to develop interactive and dynamic models of contention (McAdam, Tarrow and Tilly 2001; McAdam, Tarrow and Tilly 2008; C. Tilly 2008).
3 Financial integration in 2007 was not universal. Industrialised economies were highly integrated, and the banking crisis transmitted rapidly from one rich country to another. Middle- and low-income countries were far less exposed, and felt the effects of the financial crisis primarily through its volatile effect on commodity prices including energy and food.
4 Bangladesh Financial Express, 17 May 2015, http://print.thefinancialexpress-bd.com/2015/05/17/92935/print (accessed 13 March 2017).
5 The *Life in a Time of Food Price Volatility* project was a four-year collaboration between Oxfam; the Institute of Development Studies, University of Sussex, UK; University College Cork, Ireland; and research partners in ten focus countries: Bangladesh, Bolivia, Burkina Faso, Ethiopia, Guatemala, Indonesia, Kenya, Pakistan, Vietnam and Zambia. It was financed by the UK Department of International Development and Irish Aid. The research involved yearly return visits to twenty-three urban and rural communities in ten countries, and analysis of national and international food data.

Bibliography

Ahmed, Nazneen, and Dev Nathan. 2014. *Improving Wages and Working Conditions in the Bangladeshi Garment Sector: The Role of Horizontal and Vertical Relations*. Dhaka: Bangladesh Institute of Development Studies.

Araghi, Farshad. 2009. "Accumulation by Displacement: Global Enclosures, Food Crisis, and the Ecological Contradictions of Capitalism." *Review (Fernand Braudel Center)* 32 (1): 113–146.

Arrighi, Giovanni. 2010. *The Long Twentieth Century: Money, Power and the Origins of Our Times*. 2nd ed. London: Verso.

Badiou, Alain. 2012. *The Rebirth of History: Times of Riots and Uprisings*. Translated by Gregory Elliot. London. New York: Verso.

Bar-Yam, Yavni, Marco Lagi, and Yaneer Bar-Yam. 2015. South African Riots: Repercussion of the Global Food Crisis and US Drought. In *Conflict and Complexity: Countering Terrorism, Insurgency, Ethnic and Regional Violence*, edited by Philip vos Fellman, Yaneer Bar-Yam and Ali A. Minai: Springer.

Bassel, Leah. 2012. *Media and the Riots: A Call for Action*. London: Citizen Journalism Educational Trust.

Bellemare, Marc F. 2015. "Rising Food Prices, Food Price Volatility, and Social Unrest." *American Journal of Agricultural Economics* 97 (1):1–21. doi:0.1093/ajae/aau038.

Block, Fred L., and Margaret R. Somers. 2014. *The Power of Market Fundamentalism: Karl Polanyi's Critique*. Cambridge, MA: Harvard University Press.

Bohstedt, John. 2010. *The Politics of Provisions: Food Riots, Moral Economy and Market Transition in England c.1550–1850*. Farnham: Ashgate.

Bohstedt, John. 2016. "Food Riots and the Politics of Provisions From Early Modern Europe and China to the Food Crisis of 2008." *The Journal of Peasant Studies* 43 (5):1035–1067. doi:10.1080/03066150.2016.1170009.

Brinkman, Henk-Jan, and Cullen S. Hendrix. 2011. "Food Insecurity and Violent Conflict: Causes, Consequences and Addressing the Challenges." *Occasional Paper 24*. Rome: World Food Programme.

Burawoy, Michael. 2015. "Facing an Unequal World." *Current Sociology* 63 (1):5–34.

Bush, Ray, and Giuliano Martiniello. 2017. "Food Riots and Protest: Agrarian Modernizations and Structural Crises." *World Development* 91:193–207.

Chatterjee, Partha. 2004. *The Politics of the Governed: Reflections on Popular Politics in Most of the World.* New York: Colombia University Press.

Clapp, Jennifer. 2012. *Hunger in the Balance:* The *New Politics of International Food Aid.* 1st ed. Ithaca, NY: Cornell University Press.

Cooper, Charlie. 2012. "Understanding the English Riots of 2011: 'mindless criminality' or Youth 'Mekin Histri' in Austerity Britain?" *Youth and Policy* 109:1–26.

Das, Veena. 1995. *Critical Events: An Anthropological Perspective on Contemporary India.* Delhi: Oxford University Press.

Dawe, David, and Cristian Morales-Opazo. 2009. "How Much Did Developing Country Domestic Staple Food Prices Increase During the World Food Crisis? How Much Have They Declined?" *ESA Working Paper No 09–09*. Rome: FAO Agricultural Development Economics Division.

De Waal, Alexander. 1997. *Famine Crimes: Politics & the Disaster Relief Industry in Africa.* London: African Rights and the International African Institute in association with James Currey (Oxford) and Indiana University Press (Bloomingon, IN and Indianapolis).

Demarest, Leila. 2015. "Food Price Rises and Political Instability: Problematizing a Complex Relationship." *European Journal of Development Research* 27 (5):650–671.

Easterly, William, and Stanley Fischer. 2001. "Inflation and the Poor." *Journal of Money, Credit and Banking* 33 (2):160. doi:10.2307/2673879.

Fraser, Nancy. 1997. *Justice Interruptus: Critical Reflections on the "Postsocialist" Condition:* London and New York: Routledge.

Friedmann, Harriet. 1993. "The Political Economy of Food: A Global Crisis." *New Left Review* 197:29–57.

Gandelman, Nestor, and Ruben Hernandez-Murillo. 2009. *The Impact of Inflation and Unemployment on Subjective Personal and Country Evaluations*. Rochester, NY: Social Science Research Network.

Ghosh, Jayati. 2010. "The Unnatural Coupling: Food and Global Finance." *Journal of Agrarian Change* 10 (1):72–86. doi:10.1111/j.1471–0366.2009.00249.x.

Hendrix, Cullen, and Henk-Jan Brinkman. 2013. "Food Insecurity and Conflict Dynamics: Causal Linkages and Complex Feedbacks." *Stability: International Journal of Security and Development* 2 (2):26.

Hendrix, Cullen S., and Stephan Haggard. 2015. "Global Food Prices, Regime Type, and Urban Unrest in the Developing World." *Journal of Peace Research* 52 (2):143–157. doi:10.1177/0022343314561599.

Hickey, Sam. 2013. "Thinking About the Politics of Inclusive Development: Towards a Relational Approach." *ESID Working Paper Series*. Manchester: Global Development Institute.

Holt-Giménez, Eric. 2008. "The World Food Crisis: What Is Behind It and What We Can Do About It." *Food First Policy Brief 16*. Oakland, CA; Institute for Food and Development Policy.

Hossain, Naomi, Richard King, and Alexandra Kelbert. 2013. *Squeezed: Life in a Time of Food Price Volatility, Year 1 Results*. Brighton and Oxford: Institute of Development Studies and Oxfam.

Hossain, Naomi, Richard King, Alexandra Kelbert, Patta Scott-Villiers, and Nick Chisholm. 2015. *Delicious, Disgusting, Dangerous: Eating in a Time of Food Price Volatility*. Brighton: Institute of Development Studies.

IMF. 2007. "World Economic Outlook (April 2007): Spillovers and Cycles in the Global Economy." *World Economic and Financial Surveys*. Washington, DC: International Monetary Fund.

Jayadev, Arjun. 2008. "The Class Content of Preferences Towards Anti-Inflation and Anti-Unemployment Policies." *International Review of Applied Economics* 22 (2):161–172. doi: 10.1080/02692170701880643.

Lagi, M., K. Z. Bertrand, and Y. Bar-Yam. 2011. *The Food Crises and Political Instability in North Africa and the Middle East*. SSRN http://ssrn.com/abstract=1910031

Lentz, Erin C., and Christopher B. Barrett. 2013. "The Economics and Nutritional Impacts of Food Assistance Policies and Programmes." *ESA Working Paper 13–04*. Rome: FAO Agricultural Development Economics Division.

McAdam, D., S. G. Tarrow, and C. Tilly. 2001. *Dynamics of Contention*. Cambridge; New York: Cambridge University Press.

McAdam, D., S. G. Tarrow, and C. Tilly. 2008. "Methods for Measuring Mechanisms of Contention." *Qualitative Sociology* 31: 307–331.

McMichael, Philip. 2009a. "A Food Regime Analysis of the 'world food crisis'." *Agriculture and Human Values* 26 (4) (281–295).

McMichael, Philip. 2009b. "A Food Regime Genealogy." *Journal of Peasant Studies* 36 (1):139–169.

McMichael, Philip. 2010. "The World Food Crisis in Historical Perspective." In *Agriculture and Food in Crisis: Conflict, Resistance and Renewal*, edited by Fred Magdoff and Brian Tokar, 51–68. New York: Monthly Review Press.

Moore, Jason. 2014. "Cheap Food & Bad Climate." www.researchgate.net/profile/Jason_Moore3/publication/264554597_Cheap_Food__Bad_Climate_From_Surplus_Value_to_Negative-Value_in_the_Capitalist_World-Ecology/links/53e664bd0cf25d674e9ff5dc.pdf

Moore, Jason W. 2015. *Capitalism in the Web of Life: Ecology and the Accumulation of Capital*. London: Verso.

O'Brien, Thomas. 2012. "Food Riots as Representations of Insecurity: Examining the Relationship Between Contentious Politics and Human Security." *Conflict, Security & Development* 12 (1):31–49.

Ortiz, Isabel, Sara Burke, Mohamed Berrada, and Hernán Cortés. 2013. *World Protests 2006–2013*. New York: Friedrich Ehbert Stiftung.

Patel, Raj. 2012. *Stuffed and Starved: The Hidden Battle for the World Food System*. 2nd Revised Expanded ed. Brooklyn, NY: Melville House.

Patel, Raj, and Philip McMichael. 2009. "A Political Economy of the Food Riot." *Review. Fernand Braudel Center* XXXII (1):9–36.

Polanyi, Karl. 1957. *The Great Transformation: The Political and Economic Origins of Our Time*. Boston, MA: Beacon Press.

Rashid, Mamunur, Bayazid Hasan, and Naomi Hossain. 2012. "Bangladesh: Pathways and Impacts of the Global Economic Shocks." In *Living Through Crises: How the Food, Fuel and Financial Shocks Affect the Poor*, edited by Rasmus Heltberg, Naomi Hossain and Anna Reva. Washington, DC: World Bank.

Scheve, Kenneth. 2004. "Public Inflation Aversion and the Political Economy of Macroeconomic Policymaking." *International Organization* 58 (1): 1–34. doi: 10.1017/S0020818304581018.

Scott-Villiers, Patta, Nick Chisholm, Alexandra Wanjiku Kelbert, and Naomi Hossain. 2016. *Precarious Lives: Food, Work and Care After the Global Food Crisis*. Institute of Development Studies. https://opendocs.ids.ac.uk/opendocs/handle/123456789/12190

Standing, Guy. 2011. *The Precariat: The New Dangerous Class*. London: Bloomsbury.

Standing, Guy. 2016. *The Corruption of Capitalism: Why Rentiers Thrive and Work Does Not Pay*. London: Biteback Publishing.

Taylor, Lynne. 1996. "Food Riots Revisited." *Journal of Social History*. 30 (2): 483–496.

Thompson, Edward P. 1971. "The Moral Economy of the English Crowd in the Eighteenth Century." *Past and Present* 50:76–136.

Tilly, Charles. 1975. "Food Supply and Public Order in Modern Europe." In *The Formation of National States in Western Europe*, edited by Charles Tilly, 380–455. Princeton, NJ: Princeton University Press.

Tilly, Charles. 2008. *Contentious Performances*. Cambridge: Cambridge University Press.

Tilly, Charles, and Sidney Tarrow. 2015. *Contentious Politics*. 2nd ed. Oxford: Oxford University Press.

Tilly, Louise A. 1971. "The Food Riot as a Form of Political Conflict in France." *The Journal of Interdisciplinary History* 2 (1): 23–57. doi:10.2307/202442.

UNCTAD. 2015. *The State of Commodity Dependence 2014*. New York and Geneva: United Nations Conference on Trade and Development.

Walton, John K., and David Seddon. 2008. *Free Markets & Food Riots: The Politics of Global Adjustment*. 2nd ed. Oxford: Wiley-Blackwell.

2 A world in protest

Sara Burke

Introduction

During the time period covered by the *Food Riots and Food Rights* project, 2007–2012, the world registered a dramatic eruption of protests on many issues, including food. The protests were largely non-violent, but during these years episodes reported as "riots" and "violent protests" also occurred. These spiked to historic levels in 2008, 2011 and 2012 in response to price hikes for basic commodities necessary to life: fuel, water and especially food. Protests for subsistence have not been well understood, as they were not well covered in international news due to systematic biases in professional media organizations, which tend to ignore non-violent protests – or the nonviolent parts of protests – and also fail to distinguish between acts that harm lives and those that harm property, as well as labeling "violent" both riots and protests violently repressed by the state.

This chapter aims to show how the historical eruption of protests throughout the period can be understood as a response to the stormy political economy of the time.[1] To many critics of the dominant economic paradigm, that time is best understood as the forseeable outcome of neoliberalism, what classical political economists Gérard Dúmenil and Dominique Lévy's 2011 book *The Crisis of Neoliberalism* characterizes as a globalized system of "free-market economics, free trade, and the free mobility of capital", literally a "new liberalism" and corresponding social order in which financialization propels a managerial apparatus that disciplines labor with greater and greater intensity (Dúmenil and Lévy 2011: 35). "Neoliberalism" as a term of art, is typically counterposed to "globalization". Proponents of inclusive, managed globalization have approached the period with a critique of its policy paradigm, the so-called "Washington Consensus", which Keynesian Joseph Stiglitz characterized in his 2002 book *Globalization and its Discontents* as the point at which growth, privatization and liberalization became ends in themselves, rather than means toward a more equitable distribution that benefits even the poor (Stiglitz 2002). Neoclassical economist John Williamson, originator of the term "Washington Consensus", frequently complained that the set of ten policies he felt were widely held in Washington based institutions like the IMF, World Bank and U.S. Treasury in 1989–1990 (e.g. fiscal discipline, reorientation of public expenditures, tax reform, financial liberalization, unified and competitive exchange rates, trade liberalization, openness to direct foreign investment, privatization, deregulation and secure property rights) were good policies for Latin America at that time, but had been subsequently expanded by both proponents and critics into a generalized policy prescription he had not originally intended, although he didn't necessarily disagree with it (Williamson 2004: 1, 22).

Whether classical, neoclassical or Keynesian in theoretical orientation[2] and regardless of their differing perspectives, these economists agree on the significance of the period's economic growth and the corresponding role of liberalization, financialization and deregulation in shaping the political economy of the time. Two economic events of great magnitude in the commodities and financial markets took place that were driven by these very forces. Like storms rolling in from opposite horizons in 2007–2008, historically high commodity prices hit the global economy just as mortgage-based finance began its historic collapse. Flaunted profits and conspicuous consumption, which accompanied both the commodities boom and the financial crisis, enraged populations, which believed a tacit contract between citizens and state had been broken when basic economic needs were priced out of reach (Hossain *et al.* 2014). This crisis of the "moral economy" affected not only food and fuel protests in 2007–2008 and 2011–2012, but also helps us understand why one of the key demands coming out of the so-called Arab Spring, the Movement of the Squares in Europe and the Occupy movement that began on Wall Street and spread to many cities was the demand for "real democracy" in the face of unaccountable governments.

To better examine these crises in the political and moral economy, we will draw from a data set on protests and political participation in the period 2006–2013 that was created by the author and a multilingual research team between 2013 and the present, outlining the main findings from this global study of protests.[3] This data set, which coincidentally overlaps 2007–2012 by one year at the beginning and end of the period, is also the basis of an analysis of 843 protest events[4] appearing in more than 500 news sources available on the internet, and in 84 countries representing more than 90 per cent of the world population (Ortiz et al. 2013). It takes note of main grievances and demands, who and how many were protesting and where, what methods/repertoires they used, who their opponents or targets were and what resulted from the episode of protests – not only in terms of meeting the goals of a campaign, but also in terms of negative or positive unintended consequences such as repression on the one hand, or the creation of new movements, groups or political parties on the other. The main phases in the political economy of the time break down into three parts of roughly two-year duration each – boom and bust (2007–2008), stimulus (2009–2010) and contraction (2011–2012). After first outlining main and relevant findings from the protests study, we will analyze the emergence of protest episodes during the three phases to better understand the politics of subsistence protests and their relationship to other protests at the time.

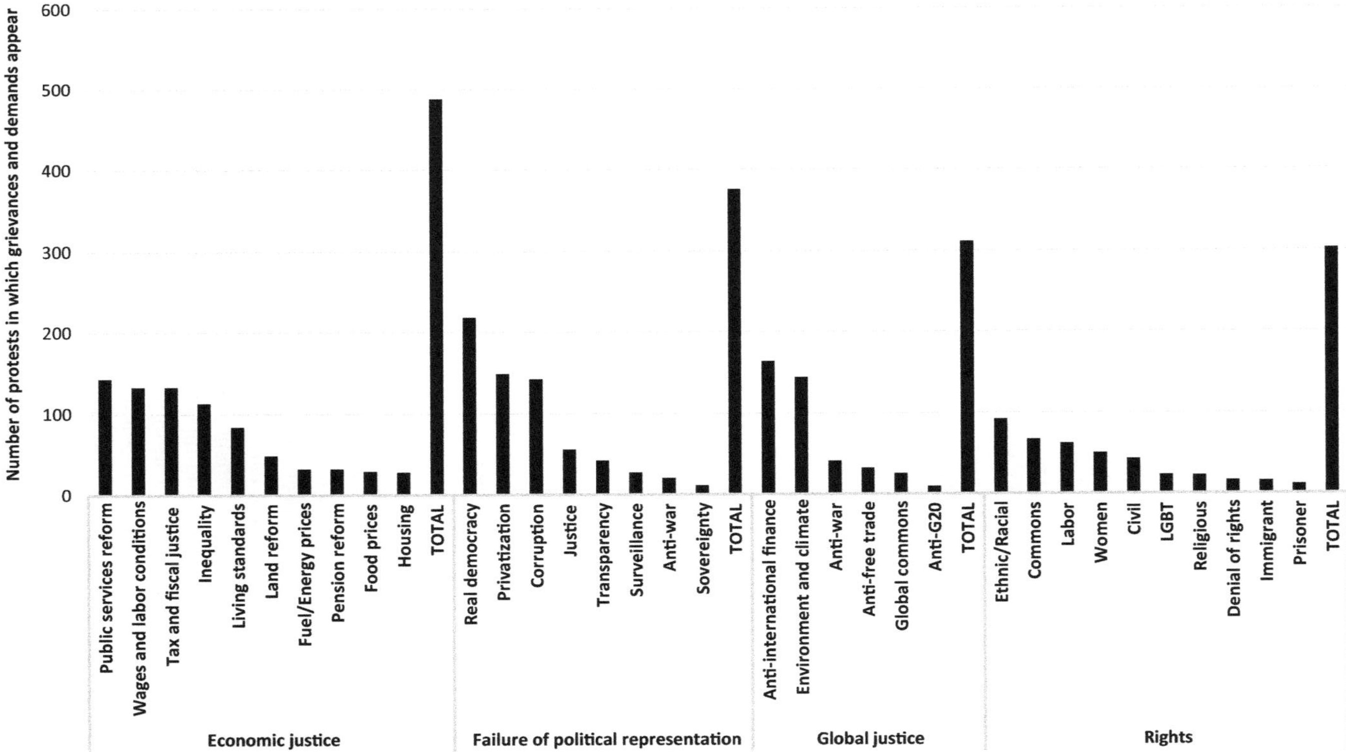

Grievances and demands grouped thematically. (Note: since multiple grievances and demands can overlap within a protest event, totals for each thematic group do not add up to the total listed for each grievance.)

Figure 2.1 Grievances and demands

Source: Global Platform on Participation and Protest data, 2014

World protests (2006–2013)

What becomes clear when we look at trends between 2006–2013 is that protests and protest movements were increasing in every geographic region and in every country income group (Ortiz *et al.* 2013: 12–13; Carothers and Youngs 2015: 22–23). The leading cause of all protests was a cluster of grievances against austerity policies and for economic justice that included demands to reform public services and pensions, create good jobs and better labour conditions, make tax collection and fiscal spending progressive, reduce or eliminate inequality, alleviate low-living standards, enact land reform and ensure affordable food, energy and housing (Figure 2.1). The key demand in a second cluster of grievances relating to failure of political representation and political systems was for "real democracy". It was also the single demand found most in world protests of the period and was counterposed to the actually-existing, corrupt and dysfunctional representative democracies that were increasingly faulted during this period for serving elites and private interests and for failing to stop growing economic inequality. Along with protests against corporate influence on government, deregulation, privatization and corruption, the demand for real democracy was coming from every kind of political system, not only authoritarian administrations but also representative democracies, old and new.

A third cluster, of globally organized – or globally aspirational – protests, was less coherent than the other clusters, but encompassed rising environmental and climate movements, some aspects of Occupy and the Movement of the Squares, some indigenous movements and the ongoing alter-globalization movement against free trade and multinational agribusiness, as well as the burgeoning open internet/internet commons movement of whistleblowers, leakers and anarchist-hackers.

Whistleblowers, leakers and anarchist-hackers

This movement includes actors as diverse as Samuel Provance, who exposed U.S. military torture in Abu Ghraib prison in Iraq in 2006, as well as whistleblowers/leakers Mark Klein, Thomas Drake and Edward Snowden, who each revealed aspects of U.S National Security Administration domestic spying (Warren and Dirksen 2014: 72–74) and military intelligence analyst U.S. Army Cpl. Bradley Manning,[5] who leaked 750,000 classified and sensitive military and diplomatic documents in early 2010 to raise awareness of the human costs of war. These leaks, which came to be known as the "Iraq War Logs" (Wikileaks 2010a) and "Afghan War Diary" (Wikileaks 2010b), were published on the website Wikileaks by self-proclaimed "cypherpunk" Julian Assange (Assange et al. 2012). While each of these agents acted independently, the hacktivism movement coalesced around the group Anonymous, which – among a variety of vigilante actions in different countries – protested the authorities' mishandling of a rape and sexting case in Nova Scotia, Canada, resulting in the subsequent suicide of victim Rehtaeh Parsons. Anonymous hackers determined the identies of the perpetrators and threatened to expose them if relevant government and law enforcement authorities did not hold them accountable to the rule of law (CTV News Atlantic 2013; Slifer 2013; Kotz 2013; Griner 2014). The activities of the movement contributed to an upsurge of interest in acts of solidarity such as mutual aid and building a global information commons, but in the case of the Wikileaks and the revelations of Edward Snowden, also inadvertently acted as catalysts to world events that became violent or were met with violence, beginning with Wikileaks' impact on the Arab Spring. In Tunisia, the population was already roiled by these leaks when Wikileaks published a U.S. dispatch with damning footage of their leaders' opulence and decadence within weeks of when the revolution began (we explore this further in the section on *Contraction*).

A fourth cluster of grievances and demands framed in the language of rights (human, civil, racial, religious, labour, women's, indigenous, LGBT rights and so on), also emerges. During the period we are concerned with, especially 2011–2012, demands for rights seem to be fewer in number than the other demand clusters. An exception is demands for labor rights and "subsistence rights" to food and fuel; this is most likely because both encompass economic as well as civil/political demands. Labor does so for both wages and freedom of association. Demands for subsistence are made for the implied contract between people and government that "the price of basic commodities such as food and fuel does not rise beyond the reach of 'the ordinary person'" (Musembi and Scott-Villiers 2014: 36).

What the majority of protests of this period have in common – regardless of where they took place, where the governments they challenged fell on the political spectrum or whether protesters were violent or not – is that they reveal failures of economic and social development and a demand for democracy that is capable of meeting people's economic needs with effective policies and laws. In the aftermath of a powerful economic crisis that reverberated from the core of the global economy to the periphery, and with terror and armed conflict going on in many parts of the globe, the repression of protests amidst growing economic and social inequalities contributed strongly to a loss of faith in official political processes and usual political actors.

A closer look at who was protesting, and how, reveals not only the traditional trade unions and activists from NGOs, but also middle-class protesters of all ages, from students to pensioners, who did not consider themselves activists but nevertheless did not trust the current political and economic system. Beginning in 2011, they joined movements in the streets in large numbers, not only in sanctioned marches and rallies, but in a new framework of protest that included direct actions such as "riots", road blockages and occupations of city streets and squares or private businesses like banks (Figure 2.2).

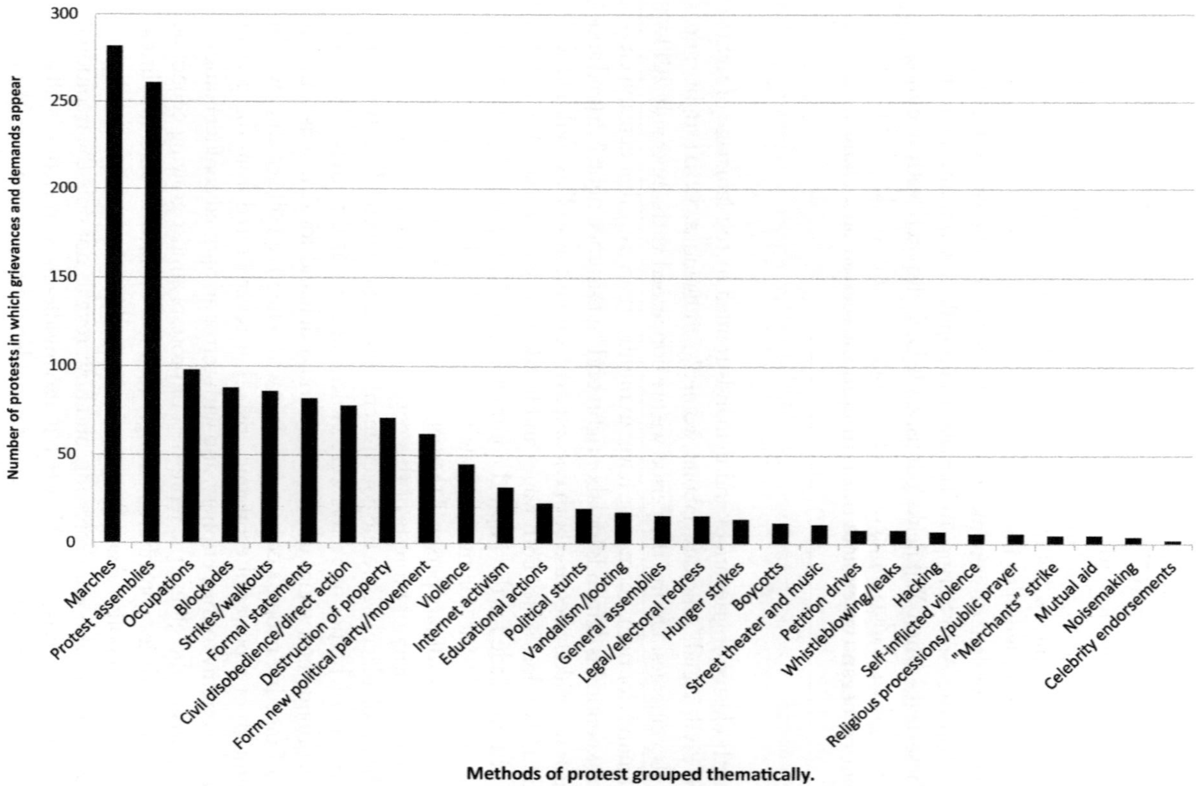

Figure 2.2 Methods of protest

Source: Global Platform on Participation and Protest data, 2014

Contrary to public perception about rioting, actual protester violence was a small part of overall protests, although a not-insignificant number of violent protests involved self-harm, desperate methods such as hunger strikes or self-immolation of the kind that sparked the protests in 2010 leading to revolution in Tunisia. Nevertheless, not only riots and other violent acts, but more than half of protest episodes globally experienced some sort of repression in terms of arrests, injuries or deaths at the hands of authorities, or subsequent surveillance of suspected protesters and groups, which both governments and private corporations carried out. Of reported riots, almost half were in low income countries, mostly prompted by spikes in food and energy prices (Ortiz *et al.* 2013: 12). We turn to this issue in more detail in the next section.

The data informing the Global Platform on Participation and Protest does not represent an algorithm-driven approach, nor does it attempt an exhaustive "big-data scooping" of protests, which typically still relies on rough categories of actors' behaviors to encode information.[6] While the field of data analytics is evolving rapidly, as of late 2014 none of the few public, data-driven resources on protests at the global level – such as the Global Database of Events Language and Tone (Leetaru 2014), which predicts events of protest and conflict by analyzing people's feelings as expressed online – has allowed researchers to "drill down" to the level of individual, citable sources or to trace complex processes in which multiple new actors made demands on various local, national and international opponents, sometimes in multi-year campaigns (Barrada 2014). The quality of the "big data" available as of 2013–2014 was therefore not sufficient to explain why and how the new waves of protests during this period had emerged.

What the Global Platform on Participation and Protest's hand-curated approach captures best are months- or years-long episodes of protest whose impact was both newsworthy (in other words, it was reported in major national or international media) and for which strong evidence of grievances and demands is documented in protesters' own words in local or alternative news sources or social media. This dual-sourced approach was an attempt to guard against known bias in mainstream news coverage by querying not only international but also local news and activist sources available on the internet (Musembi and Scott-Villiers 2014). In hindsight in early 2017, another bias emerges for which social scientists, especially those studying social movements, are only beginning to correct. This has been a tendency to focus on left-wing social movements oriented to human rights and social justice at the expense of greater examination of the right-wing movements for denial of rights and nationalist protectionism that have formed before our eyes during this period, and which now overshadow many of the progressive movements that arose in 2010–2011.[7]

The politics that arise from the indignation of having to protest for something as basic as food flows directly from the impact on populations when the implicit contract between people and governments to keep basic economic needs within reach – a key tenet of the moral economy – has been broken (Hossain *et al.* 2014; Musembi and Scott-Villiers 2014). This is a ritualized and indirect politics that must not be taken uncritically at face value. Reports of violence that tar and discredit legitimate protests must be scrutinized: which agents incited the violence, and how? The targets of food protests are almost always national governments, but international forces working on those governments through trade deals, tax competition, debt servicing and financial regulation should not escape responsibility. Why? This entire constellation of actors – the protesters, national governments, international organizations, national and international media – have a role in determining whether protests for subsistence are treated as legitimate grievances requiring governmental or intergovernmental response, or as conflicts to be put down (Burke 2015).

Boom and bust (2007–2008)[8]

In the early 2000s, financialization was rapidly increasing in every region, but especially in high-income countries where falling labor share of GDP forced workers to take on personal debt to finance living standards, much of which was with imported goods from developing countries, in the face of falling real wages and rising inequality (Duménil and Lévy 2011: 179; Dünhaupt 2013; Gindin 2013; Palley 2007). Non-financial firms contributed to the trend, embracing leveraged financialization of major portions of their profits in lieu of traditional reinvestment in the firm and payouts to shareholders (Duménil and Lévy 2011: 126, Rude 2005: 86). Alongside these developments, the historic commodities boom that began in 2000 fueled economic growth in developing countries up until the 2007–2008 economic crisis, which reverberated from the core of the global economy in 2007–2008 to the periphery in 2008–2009 and, for a few years, effectively stopped the commodities boom in its tracks (Erten and Ocampo 2012).

Prior to the crash, manufacturing and trade soared in the BRICS countries, particularly China and India, where the long economic expansions, as of 2007, were compared by IMF to the post-World War II experience of fast growing European economies, Japan and the newly industrializing economies in Asia (IMF 2007: 173). Following the 1997 currency crisis, Asian economies that borrowed from IMF had bristled at what they considered harsh and excessive conditionality by the institution. Consequently, during the 2000 commodity boom, countries throughout Asia accumulated trillions of U.S. dollars in foreign exchange reserves as "self-insurance" so that they would never again have to appeal to the IMF for bailouts (Lin and Treichel 2012: 24). These reserves contributed to global imbalances that, alongside housing bubbles in the United States and Europe and a lack of regulation in the financial sector at all levels, led to volatility in the global economy (Obstfeld and Rogoff 2009). These were arguably foreseeable outcomes of contradictions in the neoliberal globalization project on which economic arrangements were premised: free trade without free immigration, growth and deregulation without universal social protection, financialization exacerbating inequalities within and between countries and groups.

By mid-2007, as commodity prices had skyrocketed, a financial crisis was stirring quietly in the housing markets of the U.S. and Europe (Pisani-Ferry and Sapir 2009: 9). It continued to build through the autumn and winter in spite of a discreet series of coordinated central bank swaps aiming to curtail the toxic effects of exposure to exotic financial instruments at the heart of sub-prime housing loans (Rude 2010). By late September 2008, sub-prime's quiet turmoil had erupted full-force into the public sphere, with failed banks, frozen credit markets and political upheaval on both sides of the Atlantic.

The immediate aftermath of the banking collapse focused on damage control. In Ireland, this served banking interests: on October 1, 2008 without any public debate, the government guaranteed the debt of the six largest financial institutions and just a few months later announced public spending cuts. In Iceland, protesters ensured the crisis response was in the public interest: banks were nationalized on October 6, 2008; IMF and European Union loans were procured; and the Prime Minister was forced to resign and make way for a center-left government that formed in early 2009 on a vow to rebuild the economy (Guillén 2015).

From the eruption of this trans-Atlantic crisis in mid-September 2008, hardly a month passed before the regional crisis raised global alarm. The result was an upgrade to the Group of 20 (G20) from a gathering of finance ministers and central bankers to a summit of heads of state. By the time U.S. President George W. Bush welcomed anxious G20 presidents and prime ministers to Washington, D.C. in mid-November 2008, the global financial system had lost capacity to take even the simplest steps to convert corporate savings into investment, home building, personal consumption or development. In contrition, the G20 issued a declaration promising to address root causes, take coordinated actions and reform financial markets through transparency, accountability, sound regulation, prudential oversight, risk management, integrity (of markets) and international cooperation (G20 2008).

The first order effects of this crisis, which devastated financial markets in the United States, United Kingdom and European Union, were largely contained by the quick G20 interventions to segregate commercial and investment banking, exchange information for better international tax cooperation and discuss principles for sound executive compensation, along with other measures to curb short-term excessive-risk taking (Rude 2010: 1–3). Alarm catalyzed cooperation among governments, and between governments and finance, but since the crisis unfortunately did not end with financial market recovery, its second- and third-order effects began to hit ordinary citizens in the U.S., U.K. and E.U. in the form of collapsing savings and pensions, falling incomes, unemployment, foreclosures, homelessness and lack of access to food, clean water and medical care (UN 2011: 65–66).

Middle-income and low-income developing countries were also affected as commodity price indices first spiked and then became radically volatile over the next few years. Trade was hit hard; slower growth dampened manufacturing and export prospects. People living in lower middle- and low-income countries faced multidimensional poverty, few decent work opportunities and economically compelled migration, on top of rising inequalities in many countries (Moghadam, Cottarelli and Blachard 2011: 8–9; World Bank 2011: 11). For large parts of the world population, hopeful development outcomes and peaceful globalization premised on the United Nations' (UN) 2000 Millennium Development Goals and the subsequent "Monterrey Consensus" of all UN heads of state in 2002, were hamstrung by unmet promises of official development aid from high income countries, and the crisis in G20 countries made it all the more likely those promises would remain unfulfilled. Poverty in developing countries persisted despite growth fueled by the commodities boom in the first part of the decade. Corruption in governments reinforced that trend, as did sophisticated tax avoidance by multinational companies, which deprived developing countries of resources needed to provide essential social services for their populations.

Almost two years passed between the emergence of the crisis in 2007 – which built slowly and in an almost hidden way throughout 2007 and until mid-2008, fueled in part by historically high commodity prices – and the crisis' explosion into the public sphere when Lehman Brothers declared bankruptcy on September 15, 2008.[9] For this time period, 2007–2008, there are ninety-eight protest episodes in the GPPP database that shed considerable light on the main issues driving protests in the build-up to the economic crash. They relate primarily to economic grievances, rights and failures of political representation.

Discontent around economic grievances found sharp expression in Egypt in 2007, when, after a year-long series of labor actions, striking textile workers from the largest public sector textile company in the country occupied their factory and won major concessions from the government, including bonuses, corruption investigations and the participation of the government in direct negotiations for the first time in years (GPPP 2017b, BBC News 2010, Bisgaard-Church 2007; Kenyon 2007). Internationally notable labor struggles also occurred among low-skilled workers in Bangladesh (GPPP 2017c) and among highly skilled workers in Australia (GPPP 2017d). The data from 2007–2008 also capture burgeoning discontent in countries that had already begun to implement austerity, such as Greece, where hundreds of thousands protested privatization, wage caps, pension restructuring and the shutdown of public media that were part of a reform of ailing public institutions beginning in early 2007 (GPPP 2017e, f).

In hindsight, mobilizations of 2007–2008 also revealed growing social divisions in the interpretation of *rights*, such as in Poland where opposing groups demonstrated for and against tightening abortion laws (GPPP 2017g), and in Tunisia, where workers' movements were joined by unemployed youth and students, staging multiple strikes to protest their right to affordable food, an unjust employment policy and corruption (GPPP 2017h). In China (GPPP 2017i) and in indigenous communities throughout the Americas (GPPP 2017j), there were large and ongoing environmental rights protests. There were also protests against Islamic extremists in Algeria (GPPP 2017k). There were calls for a pan-European movement for immigration restrictions in Denmark, Germany and the U.K. (GPPP 2017l), while in largely Muslim Malaysia, Hindus protested religious discrimination (GPPP 2017m). The global anti-sexism protest group, FEMEN, also got its start in Ukraine among anti-sex trafficking activists in 2008 (GPPP 2017n).

Some protests presented nascent challenges to the state, as when Irish citizens protested the illegitimate socialization of debt (GPPP 2017o) and Italians protested government handling of the economic recession (GPPP 2017p). Additional protests, unified by grievances against corruption, unaccountability and failures of representation, are found in 2007–2008 in Bangladesh, Bolivia, Brazil, Cameroon, Kenya, Madagascar, Nigeria, Pakistan, the Philippines, Slovenia and Tunisia (GPPP 2017q). Icelanders protesting the financial crisis went all the way and toppled their government (GPPP 2014r), while Anonymous and the internet culture of hackers, leakers and whistleblowers in high-income countries emerged as a surprising force for exposing government and private corruption (GPPP 2017s).

There is also evidence of food riots, which are anchored in both rights and failure of political representation, in every region. While they occurred primarily in lower middle-income and low-income countries, there is evidence that in 2007–2008 food riots were reported in every country income group, even high-income countries like the United Arab Emirates, due to immigrant workers' grievances about wages and prices (GPPP 2017t). The Global Platform search for protests related to food prices lists 19 episodes during 2007–2008, in Argentina, Bangladesh, Burkina Faso, Cameroon, Egypt, Ethiopia, Haiti, Côte d'Ivoire, Kenya, Mauritania, Morocco, Mozambique, Nicaragua, Nigeria, Peru, Senegal, Somalia and Uzbekistan (GPPP 2017u).

The high degree of global integration that food and agricultural systems had attained after years of structural adjustment in the 1980s and 1990s had oriented developing countries' agricultural production toward export, making them vulnerable in the 2000s to price rises due to commodity speculation and the further degradation of food supplies when food crops were substituted for biofuel production (UN 2011). Many food- and fuel-price protests of the period were directly related to the removal of subsidies and increases in regressive taxes as a result of the expansion of austerity measures, such that by the end of the period we are examining – not only in Europe but in over 100 countries, many of them low income – public expenditures were contracting (Balakrishnan, Elson and Patel 2009; UN 2011; Ortiz and Cummins 2013).

Against this backdrop, commodity prices began to rise more sharply beginning in January 2007 (Figure 2.3). By the end of that month, so-called "tortilla riots" were reported in Mexico (Table 2.1), as farmers protested price rises due to implementation of the final stages of the North American Free Trade Agreement (Schneider 2008: 41–42). In the ensuing months, commodity prices continued to climb, setting off miners' strikes and food price protests in Peru (Schneider 2008: 46–47). Finally in July, commodity prices reached a peak worldwide and food riots continued. By the end of 2007, food riots had also been reported in India (August), Morocco, Uzbekistan (September), China, Mauritania, and Senegal (November); by the end of 2008, additional riots or violent protests for affordable food had been reported in at least 22 countries (Table 2.1).

The wave of protests for food and subsistence rights emerges as the single most powerful and numerous cluster of episodes within this two-year period. Of the

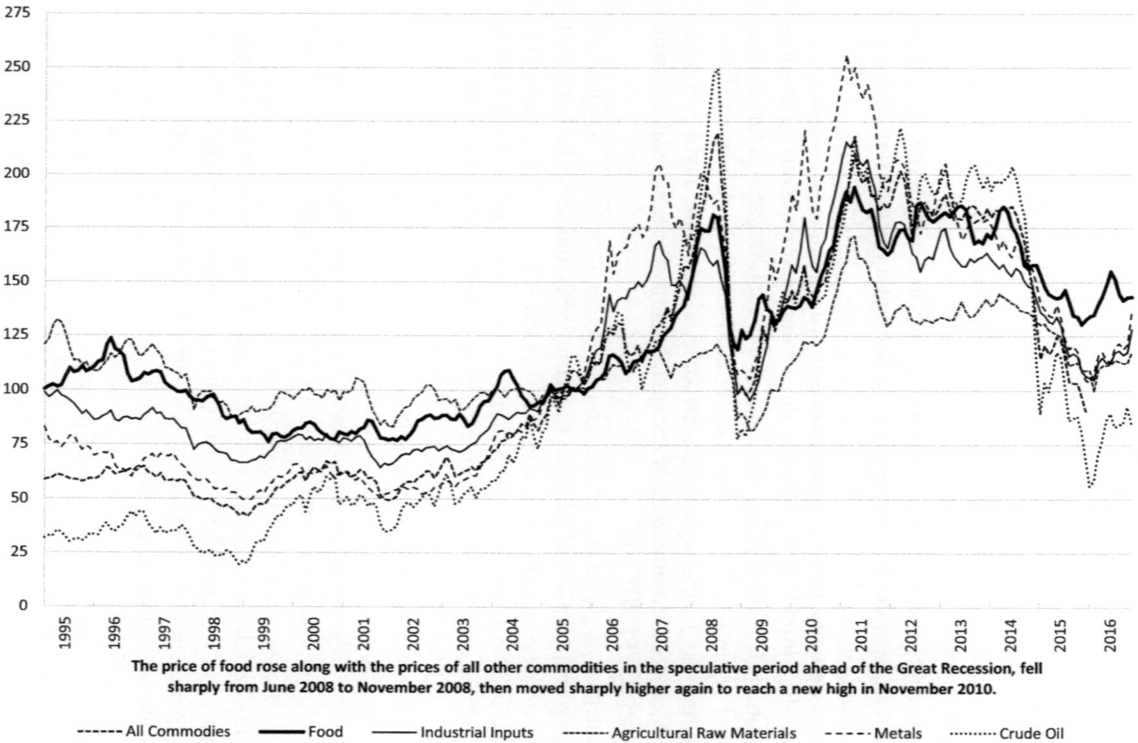

Figure 2.3 Commodity prices, 1995–2015

Note: Monthly, Indices, 2005=100

Source: International Monetary Fund

Table 2.1 Global food riots, 2007–2008

Country	*Date*
Mexico	01/01/2007
India	01/08/2007
Morocco	01/09/2007
Uzbekistan	01/09/2007
India	01/10/2007
China	01/11/2007
Mauritania	01/11/2007
Senegal	01/11/2007
Indonesia	01/01/2008
Morocco	01/02/2008
Philippines	01/02/2008
Yemen	01/02/2008
Côte d'Ivoire	01/03/2008
Senegal	01/03/2008
United Arab Emirates	01/03/2008
Bangladesh	01/04/2008
Haiti	01/04/2008
Madagascar	01/04/2008
Philippines	01/04/2008
Thailand	01/04/2008
Trinidad and Tobago	01/04/2008
Somalia	01/05/2008
Kenya	01/06/2009
Kenya	01/06/2011
Mozambique	01/12/2012
Mozambique	01/02/2013
Nicaragua	08/06/2017
Peru	01/2008–07/2007
Guinea	1/1/2007–2/1/2007
Pakistan	1/1/2008–2/1/2008
Afghanistan	2/1/2008–3/1/2008
Burkina Faso	2/1/2008–3/1/2008
Cameroon	2/1/2008–4/1/2008
Egypt	3/1/2008–4/1/2008
Ethiopia	3/1/2008–4/1/2008

98 episodes in the GPPP dataset for 2007–2008, 19 (19.4 per cent), in almost as many countries, explicitly list food prices as a primary grievance driving the protests, and characterize the incident as violent. This is in contrast to the trend over the entire data set of 2006–2013, in which only 30 episodes (6.3 per cent) were driven by food price grievances. It also contrasts with the trend from 2007–2012, in which there are 27 episodes (7.9 per cent). In other words, when we examine the whole period, riots do not predominate, but when we focus in on the roughly two-year runup to the economic crisis that began to engulf the global economy in late 2008, they stand out as the dominant response to the global economic turmoil. The riots of 2007–2008 were often not as violent as portrayed:

for example, when Al Jazeera reported in September 2007, "Morocco rolls back bread price hike: violent protests force government to withdraw 30 per cent hike in bread prices", it failed to give an accurate picture of the character of the protest, which had been organized by Moroccan Association for Human Rights as a peaceful sit-in (Al Jazeera 2007). Violence would likely have come only from the police crackdown, if at all.

More than mere media bias, this sort of mischaracterization had become a key part of the ritualized and indirect system by which people held governments to account for disregarding the "moral economy" in which governments safeguard populations against the erosion of basic economic needs, particularly food (Hossain *et al.* 2014; Musembi and Scott-Villiers 2014). Like food protests, the protests of 2011 in Tunisia, Greece, Egypt, Spain, Portugal, Italy, the U.S. and beyond were fundamentally against an immoral and unaccountable global economy and inept, aloof, corrupt or authoritarian national governments. Despite their lack of a sophisticated network, food protests became the dominant mode of agitation for accountability during the commodities boom and financial bust of 2007–2008. That is when riots in poor and developing countries were an immediate result. Three years later, as the economic crisis evolved from the over-leveraged banking sectors in North Atlantic countries before the crisis, to coordinated global stimulus post-crisis in 2009–2010, to a split between countries in the G20 on stimulus versus austerity policies mid-2010, the effects on growth, trade, financialization, unemployment and inequality set the stage for a set of sharp new agitations for accountability to arise: the Movement of the Squares, Arab Spring, and Occupy.

Stimulus (2009–2010)[10]

When the simmering financial crisis finally erupted in the autumn of 2008, G20 countries agreed on economic stimulus to boost aggregate demand to salvage GDP growth (G20 2008). The effect was primarily positive, but limited, since rehabilitated banks that sat on large cash reserves could still not be enticed to lend, and economic growth rebounded slowly when at all. Vast public resources were directed to bail out private banks considered "too big to fail", pushing the losses onto taxpayers and causing sovereign debt levels to rise, which in turn impeded economic growth (Ortiz and Cummins 2013: 37). Even if stimulus was slowing the negative social impacts of the crisis and contributing to economic turnaround in a way that technocrats could applaud, it did not feel that way to a lot of ordinary citizens of affected countries.

Developing countries that were experiencing the ongoing negative effects of the crisis on growth and trade worried that the falling fortunes of rich countries would prevent them from receiving official development assistance. Promises made at the United Nations Summit in Monterrey, Mexico in 2002 for developed countries to provide aid to developing countries amounting to 0.07 per cent of GDP per year were mostly forgotten, and those countries that kept their promises provided less aid in absolute terms as their GDPs fell to the lowest point in the first quarter of 2009 (Stiglitz 2009: 98). In response, the UN held in mid-2009 the Conference on the World Financial and Economic Crisis and its Impact on Development. The expert commission advising the conference recommended that the UN create a "Global Economic Coordination Council" to ensure that the needs of all countries, and not only the G8 or G20, would be taken into account in the crisis response (Stiglitz 2009: 87). Not surprisingly, the recommendations of the commission regarding impacts on development and multilateralism did not have much weight in the international institutions or the G20, which managed the crisis response.

Amidst speculation in early 2009 that a set of debt troubled countries were in danger of default, the E.U. ordered Ireland and Greece to reduce their budget deficits, which generated political turmoil and public anger, and it put France and Spain on notice that they should be prepared to do the same (Barber 2009). Late in 2009, Greece finally set off a crisis throughout the Eurozone when its new Socialist prime minister accused the previous government of concealing excessive borrowing; this coupled with a sharp downgrading of Greece's debt sparked concerns about the financial health of other euro area countries (Mongelli 2013: 18). Before long the debt-troubled countries were being dubbed the "PIIGS" (Portugal, Ireland, Italy, Greece, Spain) – vilified in contrast to the rising BRICS (Brazil, Russia, India, China, South Africa), and subject to more and more stringent terms of austerity in return for bailouts that did not even resolve their problems (BBC News 2010).

By mid-2010, the Euro was falling and Greece had received the first of several bailouts, although protesters stormed the Acropolis shortly after to protest the austerity measures demanded in the terms of the bailout (Smith 2010). Since 2008, the G20's initiative for regulation of big banks had stalled far short of the comprehensive, counter-cyclical policies needed to make the global financial system less volatile. By the time of the 2010 summit in Toronto, the consensus on stimulus had crumbled. Europe embraced contraction, while the U.S. and U.K. persisted with stimulus (Ortiz and Cummins 2013: 11). Shortly before year's end, Ireland revealed that the cost of its early 2009 Anglo Irish Bank bailout would push its budget deficit to unsustainable levels; the government then applied for bailout funds from the IMF and E.U. and began a new round of budgetary contraction (Whelan 2013: 13–15).

For 2009–2010, there are 92 protest episodes in the GPPP database. The majority of these concern economic needs and democratic reforms, especially the reform of public services – since austerity intensified in Greece (GPPP 2017v) and Ireland (GPPP 2017w) even while stimulus was still advocated within G20 until 2010. Other key demands were for tax justice and living wages. Protests for real democracy appear at this time in at least 17 countries, and in virtually every region (GPPP 2017x). Widening inequalities between groups in many countries and global inequalities exacerbating geopolitical tensions set the stage for protests to explode in 2011, first in Tunisia and then in Egypt, when hundreds of thousands filled their streets and squares demanding the removal of corrupt leaders, but there were still fissures, especially in Tunisia and Egypt, that had gone undetected. United Nations *Human Development Reports* and World Bank *World*

Development Reports from 2009 and 2010 forecast upbeat development prospects for both countries. Tunisia boasted high rates of internet connectivity and on the eve of the revolution may have had the highest rate of Facebook users in the world (Bachrach 2011), but its development plan to transition to a knowledge-based economy was hampered by inequality, a lack of financial inclusion and high rates of youth unemployment, on top of growing religious/secular divisions. Egypt, with its powerhouse cotton and textiles sectors and a growing financial sector, had many development projects on the books in 2010, from microfinance to airports and wind power (World Bank 2010: 29). But inequality, government corruption and the brutal repression of years of strikes by textile workers and others in the Nile Delta left the population impoverished. Both populations felt the sting of unequal societies and were angered by the lack of accountability and displays of conspicuous wealth of their leaders.

Hacks and leaks played a role in "turning up the heat" on events in the Middle East and North Africa (MENA). In Tunisia in particular, where the population had already been roiled by Wikileaks' release of the "Iraq War Diaries" and related leaks, a dispatch from the U.S. Ambassador to Tunisia went public shortly before the fateful self-immolation by college graduate/street fruit-seller Mohammed Bouazizi in December 2010. In it the ambassador describes a visit to President Ben Ali's son-in-law, whose opulent estate boasts a tiger that "consumes four chickens a day" and desserts flown in from Saint-Tropez (Bachrach 2011). Between Al Jazeera, Facebook and "Tunileaks" a site that mirrored the Tunisian-relevant postings of Wikileaks, Tunisians voraciously consumed this news of conspicuous consumption by a hated member of the elites: Bouazizi's suicidal protest was a critical spark on a powderkeg. Before 2010 had ended, the trade unions, particularly the Union Générale Tunisienne du Travail (UGTT), played a central and decisive role in brokering a mostly peaceful transition in which Ben Ali stepped down and Islamist parties began to participate in governing alongside secular parties (Yousfi 2013).

Contraction (2011–2012)[11]

The stage had been set for serious budgetary contraction following the period of coordinated stimulus from late 2008 through early 2010, when the North Atlantic countries and international financial institutions were entirely focused on bank bailouts and preventing economic spillovers, but the sharp downturn in growth of GDP, which became negative growth in the advanced economies in 2009, was felt in all parts of the world (Figure 2.4). By early 2010, the IMF had signaled a major shift in its policy recommendations toward fiscal consolidation – contraction – to be attained through cuts in health and education entitlements, eliminating subsidies and increasing tax revenues in both countries in recovery and those on shakier ground (Blanchard, Cottarelli and Viñals 2010; Cottarelli 2010). The OECD advice soon shifted accordingly, as the consensus emerged in the most advanced economies of the E.U. that the stimulus had run its course and growth had peaked (OECD 2010). A period of contraction and austerity began.

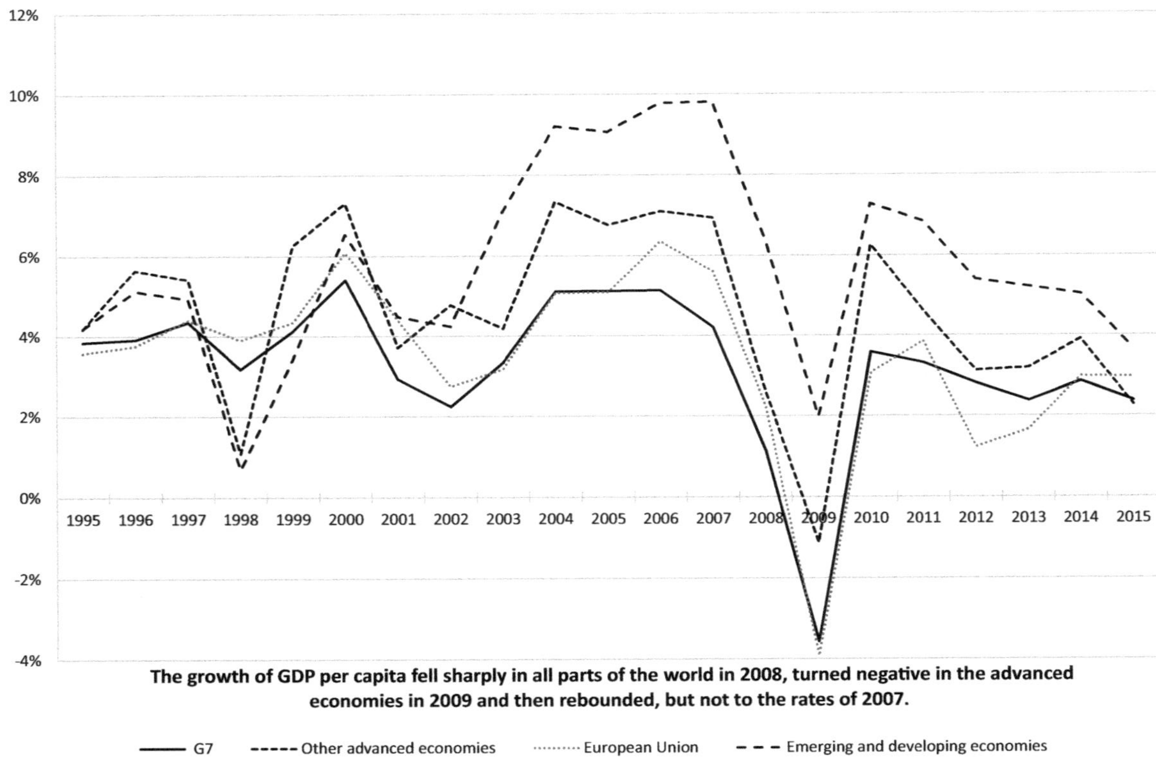

Figure 2.4 Growth of GDP per capita for selected countries, 1995–2015

Note: PPP basis

Source: International Monetary Fund

These shifts of policy outlook brought about disunity within the G20 on the need for ongoing stimulus, and enabled governments in the most advanced economies to shift blame for the crisis onto government spending and growing debt burdens, which were outcomes and not causes of the crisis (Ortiz et al. 2015: 8–9; Ní Chasaide 2013: 86–90). This shift to austerity and the growing debt crises in southern Europe continued to roil the E.U., even as GDP growth estimates for Africa, Asia and Latin America were going down and threatening development gains made over the previous decade due to falling asset prices, higher borrowing costs, capital flight and decreased exports (Li 2010: 55–57).

There were also geopolitical costs. The Tunisian and Egyptian revolutions overturned an informal arrangement between MENA countries and the European powers in which, in exchange for entry into the E.U., southern European states would vigorously enforce E.U.-wide immigration policies. In this period, immigrants in question were primarily from sub-Saharan Africa and Syria, moving through southern Europe to get to Germany or other northern E.U. states. In implicit exchange for development aid, countries along the southern Mediterranean coast, especially the Maghreb, agreed to readmit their nationals. This arrangement effectively pitted Northern Europe against Southern and had the whole of Europe against the nearest developing countries (Mouhoud 2012).

When commodity prices began to rise sharply again in early 2011 – after a steep decline from their 2008 peak – a two-year period of turbulent high prices set off another wave of food protests in at least twelve countries in sub-Saharan Africa and the MENA region (Ortiz and Cummins 2013). Other food protests in this period are likely to have gone unreported, especially in international media, overtaken in both numbers and media attention by the Tunisian and Egyptian revolutions and by the subsequent mass occupations of public squares in European and U.S. cities. These protests threw traditional political positions and alignments into question: in Europe, many Socialist Party leaders became implementers of austerity, thereby justifying widespread mistrust.

For example in Greece, when the Panhellenic Socialist Movement's (PASOK) George Papandreou became prime minister in October 2009, he discovered deep budget deficits left by the previous New Democracy administration that were in violation of the E.U.'s Stability and Growth Pact (EC 2010). Papandreou's government thus commenced to implement austerity measures and promptly appealed to the IMF and E.U. for loans in April 2010 (Guardian 2010). By 2011, he faced hundreds of thousands of demonstrators, many inspired by the revolutions that had just taken place in Tunisia and Egypt. A large proportion of these were middle class people who did not consider themselves activists but nevertheless began occupying Syntagma Square in Athens alongside trade unions and activist groups on May 5, 2011. They protested the national government's policies, IMF policies, and German domination of European Central Bank policy, because Germany insisted upon harsh austerity measures. Following the occupation of the square, these groups initiated a number of creative, community-based mutual aid experiments to fill gaps left in social services by declining government spending (Oikonomides 2013: 51–56).

Ten days after the Syntagma Square occupation began in Greece, *indignados* – ordinary citizens who were indignant over a burst housing bubble and high and rising unemployment – occupied Spain's Plaza del Sol by the hundreds of thousands. As in Greece, most people in the square did not consider themselves activists and were sharply mistrustful of the main political forces in society, but they had come to the streets in a way they had not done before (Romero 2013: 63–71; Errejón Galván 2013: 74–78). Prime Minister, José Luis Rodríguez Zapatero of the Socialist Workers Party, who won a second term in 2008, was unable throughout the period to address citizens' economic grievances and lack of trust. He resigned in April 2011, vowing not to run again. As in Greece, citizens in Spain also launched creative and sophisticated mutual aid projects to fill gaps in services (Romero 2013: 63–71).

In Ireland, which had suffered austerity since 2008, protests against taxes on water and other subsistence needs continued. People from all over the country and various political groupings boycotted water taxes and marched in the capital and in town centers bearing sharpened pikes, a traditional farmer's weapon with a rebel history. The rainy island's population was rankled at having to pay extra for water, one of Ireland's most abundant resources (Wedes 2015). Nor was political blowback from rising debt levels and austerity confined to Europe. Although the U.S. continued with stimulus after the E.U. had abandoned it, by mid-2011, the Tea Party movement in the U.S. Congress was threatening to shut down the government over routine moves to raise the debt ceiling, winning the promise of future spending cuts in exchange for lifting the ceiling just before deadline. Their brinksmanship and support for austerity helped fuel social divisions and unrest already brewing in the U.S., as occupations of bank lobbies and city streets in New York over the spring and summer of 2011 prefigured the emergence of Occupy Wall Street in September (Burke 2014).

In early 2011 in the U.K. and U.S., activists in the "Un-Cut" movement occupied bank lobbies to raise awareness of banks and other multinational corporations that paid little or no legal taxes in those countries (Tierney 2011). The movement claimed they were bad corporate citizens and deserved to be shamed. In May 2011 the so-called "May 12 Coalition" of NGO social service providers and progressive political parties and think tanks in New York City, argued for a set of policies to discipline the six biggest U.S. banks and the hedge fund managers located in the city who paid a lower rate of taxes than the average worker (May 12 Coalition 2011). As in other places, many new political actors who did not consider themselves activists began to join street demonstrations in support.

Protests continued over the U.S. "debt ceiling crisis" of the summer of 2011 until they coalesced as "Occupy Wall Street" on September 17, 2011, when activists camped out in a public-private park on Broadway in lower Manhattan for several weeks to raise awareness of the injustices of an economic system that produced unemployment, high student debt, falling wages, foreclosures, fraying social safety nets and, above all, rising income and wealth inequality (GPPP 2017y). In an essay by Judith Butler, entitled "So, What Are the Demands?" – the main question put to Occupy when it emerged as a political entity – she argues that the question, "what are the demands?" is based upon the presumption that to be legitimate, a movement must have a list of demands that it strives to have satisfied by the state (Butler 2012). Some of the demands pursued, to which Butler also refers in the essay, are central to the concerns of the movement: jobs for all, homes for all, an end to foreclosures, the repudiation of student debt, and the overarching grievance: inequality. Butler argues that no matter how comprehensive, no list of demands can communicate how they all interrelate. That is explained only by the overarching complaint, economic inequality, the "1 per cent against the 99 per cent", which emerges in Occupy and other protests of 2011 as a fundamental violation of the moral economy, when the profiteering produced and reproduced by the economic system of neoliberal globalization leaves decent, ordinary people behind.

Conclusion

The Occupy movement's greatest impact was bringing the concept of inequality to the center of public discourse. By refusing to play by the rules of protest, which required an ordered and coherent list of demands, Occupy made the claim that its demands were of a moral and ethical nature and were therefore irreducible and non-negotiable in a quest for social justice. In the years since, many have debated whether and why we care about inequality, and if we do care, what it is about inequality most concerns us.

Many economists stress pragmatic reasons; for example, because high levels of inequality might impede economic growth, create fragile economies or generate economic spillovers and contagion (Kumhof and Rancière 2011, Ostry et al. 2016). Do we then focus on income and wealth distributions at the very top 1 per cent, 0.1 per cent or 0.01 per cent? Or are we also concerned with what happens to inequality in the bottom 90 per cent or the bottom half of the income distribution, where it may be an even larger drag on growth than at the top? Do we care primarily about reducing inequalities within countries? Or like the United Nations' Sustainable Development Agenda 2030, are we also committed to a project of global development, to reduce inequalities between countries and internationally?

In a 2012 World Bank working paper, Branko Milanovic first published a chart showing the global growth incidence curve (Figure 2.5), now widely referred to as the "elephant chart", showing which segments of the global population experienced real income growth between 1988–2008 (Milanovic 2012). He has done a number of variations on the graph since then: the concept behind the graph is to show what international inequality would look like if we could line up every person in the world along the x-axis of income distribution from poorest to richest, regardless of what country they live in; the y-axis shows the income growth rate (Lakner and Milanovic 2015).

In the graph the world's poorest, mostly in Africa, line up to form the elephant's tail. Then there is the vast hump that forms the elephant's back, point A, representing the income gains of the rising middle classes in China, India, Indonesia,

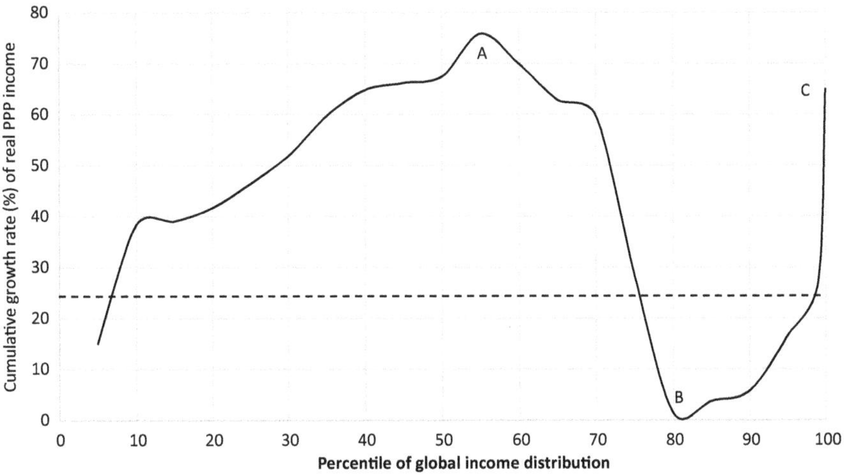

Note: y-axis displays the growth rate of the fractile average income (in 2005 PPP USD). Weighted by population. Growth incidence evaluated at ventile groups (e.g. bottom 5%); top ventile is split into top 1% and 4%

Figure 2.5 Global growth incidence curve, 1988–2008

Source: Lakner and Milanovic 2015

Thailand and other powerhouse developing countries. From that apex, the graph swings down to the lowest part of the elephant's trunk, point B, which shows the stagnant wages of the middle classes in rich countries, especially in Europe and North America. The tip of the elephant's trunk, point C, represents the global super rich. While the gains of the vast majority of the world's people in the elephant's hump indicates progress toward reducing international inequality between countries, which for over half a century has been the bulk of inequality, Milanovic cautions that even reducing inequality within every country may not have a substantial impact on global inequality and the shape of the elephant chart (Milanovic 2016).

The reasons have to do with the political and economic interests concentrated in points A, B and C of the chart and with the absence of similar powerful interests in the tail. When national borders and economies are factored into the analysis, we see how it might support the interests of the global rich (C) to play the fortunes of the rising global middle classes (A) off against the those of the falling middle classes in the old rich countries (B). This power dynamic emerges as a major roadblock to addressing inequality, especially international inequality, and has exacerbated the impact of social inequalities in many countries, contributing to rising numbers of protests on both the political left and the right and to rising numbers of global migrants in the years since 2012.

In addition, while we know quite a bit about incomes and consumption, the true depth of wealth inequality and the scale of illicit financial flows internationally are still largely hidden from view. What is increasingly not hidden is the degree of political capture of governments by wealthy interests across the ideological spectrum, making taxation and redistribution ever more difficult. This invites serious consideration of whether we have reached the limits of representative democracies to ensure widespread prosperity and social justice through laws and policies.

The initial wave of food riots in 2007–2008, which surged during the two-year period of economic crisis, held governments to account for violating the moral economy when food was scarce or unaffordable. These protests were a signal to global capitalism, a "canary in the mines" warning that the politics of subsistence must not be underestimated. Similarly, the Occupy movement and the Movement of the Squares in Europe and around the Mediterranean framed inequality as a moral and economic aberration.

These protests, different as they were from each other, had in common that they were not primarily calls for new laws or policy reforms but sharp assertions that an economic system that produces and reproduces inequality by way of its normal and predictable laws of motion violates the moral economy. In response to this violation – the idea that ordinary people should not lose their pensions, jobs, houses and prospects for a better life – Occupy refused to be silent on inequality.

Since the initial impact of the economic crisis, different parts of society in many countries have grown out of sync with each other. Consider the failure of development-aid funded NGOs to relate to street protests over food and fuel prices beginning 2007–2008 (de Brito et al. 2014) or failures of news media and pollsters in the U.K. and U.S. to accurately register public discontent with the existing political classes in 2016 in the face of populist movements for "Brexit" and U.S. demagogue Donald Trump. Groups within these societies are out of sync, and as interests collide, they have jettisoned norms of liberal democratic practice with ever greater frequency.

To understand why, it may be helpful to note that the groups in society in favor of Brexit, Trump and nationalist, protectionist movements throughout Europe and North America are aggrieved at threats, not merely to their standard of living, but to what they perceive as violations of the moral economy. It remains to be seen whether governments can be held to account for failures in development and social mobility, or if powerful interests will continue to scapegoat immigrants and refugees for a frayed moral economy. In this question, the degree of political capture of governments and institutions by rich and powerful interests is key.

Notes

1 I would like to thank my husband Christopher Rudé for his insights into the political economy of late neoliberalism.

2 Anwar Shaikh discusses these three fundamental theoretical frameworks of economics in a variety of publications. A concise explication of the differences can be found in "Heterodox Economics as a True Alternative: Going Beyond the Opposition Between 'Perfect' and 'Imperfect' Behaviors" in *New School Economic Review, Volume* 6:36–39. January 2014, New York.

3 A portion of that data can be visualized on the "Global Platform on Participation and Protest" (GPPP) http://global-platform.org evolving global documentation of contentious politics between 2006–2013 based on news and analysis openly available on the internet, still a work in progress. The data set in the GPPP, which includes new categories and further harmonization of data originally analyzed in the working paper, "World Protests 2006–2013" (Ortiz et al. 2013)

4 The main unit of analysis in "World Protests 2006–2013" is the protest event; GPPP resolved a number of events into its main unit, the protest episode, which is potentially of much longer duration. As of January 1, 2017 there were 473 protest events in the GPPP.

5 Cpl. Bradley Manning disclosed the day after being sentenced to prison that he was transgender and wished to be known thereafter as Chelsea Manning.

6 For example, the Global Database of Events Language and Tone (GDELT) purports to monitor "the world's broadcast, print, and web news from nearly every corner of every country in over 100 languages and identifies the people, locations, organizations, counts, themes, sources, emotions, counts, quotes, images and events driving our global society every second of every day" (GDELT 2017); a 2013 profile of GDELT in *Foreign Policy Magazine* characterized its output as a "massive list of important political events that have happened – more than 200 million and counting – identified by who did what to whom, when and where, drawn from news accounts and assembled entirely by software" (Keating 2013). However, its basis for classification, the "Conflict and Mediation Event Observations (CAMEO)", a coding scheme developed by political scientist Philip Schrodt for classifying actors and actions related to conflict and mediation, has an extremely limited range for categorizing protests, with only 20 verbs, of which, only seven can be said to relate to protests (Schrodt 2012).

7 The concern that social scientists' bias against studying right-wing social movements may have led to a gross underestimation of their rise in recent years was a topic of discussion at recent workshop hosted by Friedrich-Ebert-Stiftung on December 15, 2016 in Berlin on "World Protests and Political Economy". Dieter Rucht from the Wissenschaftszentrum Berlin (WZB) made the point in particular of noticing that beneath current events (e.g. the rise of the right-wing in many countries, Brexit, the election of Donald Trump as U.S. president, et cetera), three undercurrents were beginning to merge: 1) the right-wing perception of its own economic deprivation, 2) feelings of political alienation ("politicians don't care") and 3) cultural disorientation and anxiety about ways of life that are changing. The conversation acknowledged social science's failure to adequately study these groups and the need to see the proposed three undercurrents as unified.

8 The *Boom and Bust* section of this chapter references protests found in the "Global Platform on Participation and Protest" (GPPP) between January 2007 and December 2008, accessable online at http://global-platform.org/#/protests?brush=24084.329790767188%2B24109.66190634341&tags=CNT_ALL

9 Among those who locate the beginning of the financial crisis in 2007 rather than 2008 are esteemed economists George Akerlof, Olivier Blanchard, David Romer and Joseph Stiglitz, who co-edited the 2014 book, *What Have We Learned? Macroeconomic policy after the crisis*. Joseph Stiglitz begins his chapter on Lessons of the North Atlantic Crisis for Economic Theory and Policy with the locus of his analysis in "the financial crisis that began in 2007 and led to the Great Recession." He cautions therein that the wealth of historical data on economic crises and business cycles should prevent our asking, "how did this happen?" so that we might instead focus on lessons of history, including presumably our understanding of the timing of events (Stiglitz 2014).

10 The *Stimulus* section of this chapter references protests found in the "Global Platform on Participation and Protest" (GPPP) between 1/2009–2012/2010, accessable online

at http://global-platform.org/#/protests?brush=24106.971770176024%2B24131.631351710395&tags=CNT_ALL

11 The *Contraction* section of this chapter references protests found in the "Global Platform on Participation and Protest" (GPPP) between 1/2011–2012/2012, accessable online at http://global-platform.org/#/protests?brush=24131.182995682495%2B24155.842577216874&tags=CNT_ALL

References

Al Jazeera. 2007. "Morocco rolls back bread price hike." September 25, 2007. Accessed online January 27, 2017 www.aljazeera.com/news/africa/2007/09/200852514458769269.html

Assange, J., J. Appelbaum, et al. 2012. *Cypherpunks: Freedom and the Future of the Internet*. O/R Books. Accessed online February 7, 2017 https://pdf.yt/d/ekVVZgGOThtG6fXb

Bachrach, J. 2011. "WikiHistory: Did the Leaks Inspire the Arab Spring?" *World Affairs Journal* July/August 2011. Accessed online January 27, 2017 www.worldaffairsjournal.org/article/wikihistory-did-leaks-inspire-arab-spring

Balakrishnan, R., D. Elson, and R. Patel. 2009. *Rethinking Macro Economic Strategies From a Human Rights Perspective*. New York: Marymount Manhattan College/US Human Rights Network.

Barber, T. 2009. "Greece and Ireland Told to Curb Deficits." *Financial Times*. Accessed February 10, 2017 www.ft.com/content/fd3d92e4-2043-11de-a1df-00144feabdc0

Barrada, M. 2014. "Hand Collected Protest Data." *Conceptualizing Protest and Conflict: Report From an Interdisciplinary Conference Exploring How Governments and Institutions of Global Governance Can Better Respond to Contentious Politics*. FES Study, edited by S. Burke. Friedrich-Ebert-Stiftung. Accessed February 9, 2017 http://library.fes.de/pdf-files/iez/11532.pdf

BBC News. 2010. "Europe's PIGS: Country by Country." February 11, 2010, *British Broadcasting News*. Accessed February 10, 2017 http://news.bbc.co.uk/2/hi/business/8510603.stm

Bisgaard-Church, E. 2007. "Egyptian Workers Strike and Occupy Textile Factory for Better Pay, Representation, and Conditions, 2007." In *Global Nonviolent Action Database*, edited by G. Lakey. Swarthmore College. Accessed online February 10, 2017 http://nvdatabase.swarthmore.edu/content/egyptian-workers-strike-and-occupy-textile-factory-better-pay-representation-and-conditions-

Blanchard, O., C. Cottarelli, and J. Viñals. 2010. "Exiting From Crisis Intervention Policies." *International Monetary Fund paper* prepared by the Fiscal Affairs, Monetary and Capital Markets, and Research Departments. International Monetary Fund, February 4, 2010.

Burke, S. 2014. "What an Era of Global Protests Says About the Effectiveness of Human Rights as a Language to Achieve Social Change." *Sur20 International Journal of Human Rights* 11 (20): 27–34. Saõ Paulo: Conectas.

Burke, S. 2015. *Conceptualizing Protest and Conflict*. Study: July 2015. New York: Friedrich-Ebert-Stiftung.

Butler, J. 2012. "So, What Are the Demands?" *Tidal: Occupy Theory, Occupy Strategy*. https://docs.google.com/file/d/0B8k8g5Bb3BxdbTNjZVJGa1NTXy1pTk4ycE1vTkswQQ/edit?pli=1

Carothers, T., and R. Youngs. 2015. *The Complexities of Global Protests*. Washington, DC: Carnegie Endowment for International Peace. http://carnegieendowment.org/files/CP_257_Youngs-Carothers-Global_Protests_final.pdf

Cottarelli, C. 2010. "Strategies for Fiscal Consolidation in the Post-Crisis World." International Monetary Fund paper prepared by the Fiscal Affair Department. International Monetary Fund, February 4, 2010.

CTV News Atlantic. 2013. "Symbolic Protest Shows Support for Rehtaeh Parsons." *CTV News Atlantic*. Accessed February 7, 2014 http://atlantic.ctvnews.ca/symbolic-protest-shows-support-for-rehtaeh-parsons-1.1267307#ixzz2XvNUaIK7

de Brito, L., E. Chaimite, C. Pereira, L. Posse, M. Sambo, and A. Shankland. 2014. "Hunger Revolts and Citizen Strikes: Popular Protests in Mozambique, 2008–2012." Food Riots and Food Rights project report. Brighton/Nairobi: Institute of Development Studies/University of Nairobi. www.foodriots.com

Duménil, G., and D. Lévy. 2011. *The Crisis of Neoliberalism*. Cambridge, MA: Harvard University Press.

Dünhaupt, P. 2013. "The Effect of Financialization on Labor's Share of Income", *Working Paper, No. 17/2013*, Eds. Betzelt, S., Evans, T., Hein, E., Herr, H., Kronauer, M., Mahnkopf, B., Truger, A. Institute for International Political Economy, Berlin.

Errejón Galván, I. 2013. "The People United Will Never Be Defeated: The M15 Movement and the Political Crisis in Spain." *International Policy Analysis Paper*, Eds. Puschra, W. and Burke, S. February 2013. Friedrich-Ebert-Stiftung, New York.

Erten, B., and J A. Ocampo. 2012. "Super-Cycles of Commodity Prices Since the Mid-Nineteenth Century." DESA Working Paper No. 110 ST/ESA/2012/DWP/110 February 2012. New York: United Nations Department of Economic and Social Affairs.

European Commission (EC). 2010. "Report on the EDP Methodological Visits to Greece in 2010." Eurostat Directorate C: National and European Accounts. Accessed online February 17, 2017 http://ec.europa.eu/eurostat/documents/1015035/3991231/Greece-2010-methodological-visits-report.pdf

GDELT. 2017. Accessed July 13, 2017 http://www.gdeltproject.org/

Gindin, S. 2013. "Unmaking Global Capitalism." *A World to Win: Jacobin Magazine*, Issue No. 14, June 2013.

Global Platform on Participation and Protest. 2014. Accessed July 13, 2017 http://global-platform.org/#/

Global Platform on Participation and Protest (GPPP). 2017a. Accessed February 10, 2017 http://global-platform.org

Global Platform on Participation and Protest (GPPP). 2017b. "Egyptian Textile Workers Strike Over Wages." Accessed February 10, 2017 http://global-platform.org/#/protests?tags=CNT_EGY&id=95

Global Platform on Participation and Protest (GPPP). 2017c. "Bangladesh, 2007–2008, Jobs, Wages,Working Conditions Search." Accessed February 10, 2017 http://global-platform.org/#/protests?brush=24084.329790767188%2B24109.66190634341&tags=CNT_BGD%2BEJA_JWL

Global Platform on Participation and Protest (GPPP). 2017d. "NSW Electric Workers Protest Privatization." Accessed February 10, 2017 http://global-platform.org/#/protests?brush=24084.329790767188%2B24109.66190634341&tags=CNT_AU1&id=19

Global Platform on Participation and Protest (GPPP). 2017e. "Greeks Strike Over Budget." Accessed February 10, 2017 http://global-platform.org/#/protests?brush=24084.329790767188%2B24109.66190634341&tags=CNT_GRC&id=167

Global Platform on Participation and Protest (GPPP). 2017f. "Greek Journalists Protest Austerity Cuts." Accessed February 10, 2017 http://global-platform.org/#/protests?brush=24084.329790767188%2B24109.66190634341&tags=CNT_GRC&id=165

Global Platform on Participation and Protest (GPPP). 2017g. "Poles March Against Abortion." Accessed February 10, 2017 http://global-platform.org/#/protests?brush=24084.329790767188%2B24109.66190634341&tags=CNT_ALL%2BRGT_WMN&id=340

Global Platform on Participation and Protest (GPPP). 2017h. "Tunisians Protest Nepotistic Hiring." Accessed February 10, 2017 http://global-platform.org/#/protests?brush=24084.329790767188%2B24109.66190634341&tags=CNT_TUN&id=422

Global Platform on Participation and Protest (GPPP). 2017i. "Chinese Protest Location of Paraxylene Plants." Accessed February 10, 2017 http://global-platform.org/#/protests?brush=24084.329790767188%2B24109.66190634341&tags=CNT_ALL%2BFPR_COR%2BGJU_ENV&id=63

Global Platform on Participation and Protest (GPPP). 2017j. "American Indigenous Communities Demand Environmental Protections." Accessed February 10, 2017 http://global-platform.org/#/protests?brush=24084.329790767188%2B24109.66190634341&tags=CNT_ALL%2BGJU_ENV&id=463

Global Platform on Participation and Protest (GPPP). 2017k. "Algerians Protest Islamic Extremists." Accessed February 10, 2017 http://global-platform.org/#/protests?brush=24084.329790767188%2B24109.66190634341&tags=CNT_ALL&id=1

Global Platform on Participation and Protest (GPPP). 2017l. "Anti-Islam Protesters in Europe Call for Immigration Restrictions." Accessed February 10, 2017 http://global-platform.org/#/protests?brush=24084.329790767188%2B24109.66190634341&tags=CNT_ALL%2BRGT_DEN&id=89

Global Platform on Participation and Protest (GPPP). 2017m. "Malaysian Hindus Protest Discrimination." Accessed February 10, 2017 http://global-platform.org/#/protests?brush=24084.329790767188%2B24109.66190634341&tags=CNT_MYS&id=280

Global Platform on Participation and Protest (GPPP). 2017n. "FEMEN Activists Protest Patriarchy." Accessed February 10, 2017 http://global-platform.org/#/protests?brush=24084.329790767188%2B24109.66190634341&tags=CNT_ALL%2BGRV_RGT&id=142

Global Platform on Participation and Protest (GPPP). 2017o. "Irish Protest Illegitimate Debts." Accessed February 10, 2017 http://global-platform.org/#/protests?brush=24084.329790767188%2B24109.66190634341&tags=CNT_IRL&id=218

Global Platform on Participation and Protest (GPPP). 2017p. "Italians Strike Over Government Handling of Recession." Accessed February 10, 2017 http://global-platform.org/#/protests?brush=24084.329790767188%2B24109.66190634341&tags=CNT_ITA&id=238

Global Platform on Participation and Protest (GPPP). 2017q. "14 Search Results for 2007–2008 Failure of Political Representation, Transparency and Accountability, and Corruption." Accessed February 10, 2017 http://global-platform.org/#/protests?brush=24084.329790767188%2B24109.66190634341&tags=CNT_ALL%2BGRV_FPR%2BFPR_TRN%2BFPR_COR

Global Platform on Participation and Protest (GPPP). 2017r. "Icelanders Protest Financial Crisis." Accessed February 10, 2017 http://global-platform.org/#/protests?brush=24084.329790767188%2B24109.66190634341&tags=CNT_ISL&id=182

Global Platform on Participation and Protest (GPPP). 2017s. "Anonymous Takes Action Against Censorship, Corruption. Accessed February 10, 2017 http://global- rg/#/protests?brush=24084.329790767188%2B24109.66190634341&tags=CNT_ALL%2BGRP_HAK&id=144

Global Platform on Participation and Protest (GPPP). 2017t. "Emiratis Demand Higher Wages." Accessed February 10, 2017 http://global-platform.org/#/protests?brush=24084.329790767188%2B24109.66190634341&tags=CNT_United%20Arab%20Emirates%20(Dubai)&id=451

Global Platform on Participation and Protest (GPPP). 2017u. "19 Search Results for '2007–2008 Food Price Protests'" Accessed February 10, 2017 http://global-platform.org/#/protests?brush=24084.329790767188%2B24109.66190634341&tags=CNT_ALL%2BEJA_FPR

Global Platform on Participation and Protest (GPPP). 2017v. "Greeks Protest Austerity Measures, Debt Crisis." Accessed March 1, 2017 http://global-platform.org/#/protests?brush=24106.971770176024%2B24131.631351710395&tags=CNT_GRC&id=169

Global Platform on Participation and Protest (GPPP). 2017w. "Irish Workers Protest Handling of Economy." Accessed March 1, 2017 http://global-platform.org/#/protests?brush=24106.971770176024%2B24131.631351710395&tags=C

Global Platform on Participation and Protest (GPPP). 2017x. "17 Search Results for 'Real Democracy'." Accessed March 1, 2017 http://global-platform.org/#/protests?brush=24106.971770176024%2B24131.631351710395&tags=CNT_IRL&id=221

Global Platform on Participation and Protest (GPPP). 2017y. "Americans Protest Income Inequality." Accessed March 1, 2017 http://global-platform.org/#/protests?brush=24131.182995682495%2B24155.842577216874&tags=CNT_US1&id=455

Griner, A. 2014. "In Rehtaeh Parsons' Case, It's Press vs. Canadian Law." *Columbia Journalism Review*. Accessed February 9, 2017 www.cjr.org/watchdog/in_rehtaeh_parsons_case_its_pr.php

Group of 20 – G20. 2008. "Declaration of the Summit on Financial Markets and the World Economy." Washington, DC, November 15, 2008. Accessed online January 27, 2017 www.g20.utoronto.ca/2008/2008declaration1115.html The Guardian. 2010. "Greece Activates €45bn EU/IMF Loans." *The Guardian Newspaper* April 23, 2010. Accessed online March 1, 2017 www.theguardian.com/business/2010/apr/23/greece-activates-eu-imf-loans

Guillén, M. 2015. "The Global Economic & Financial Crisis: A Timeline." Working Paper of The Lauder Institute, Wharton Arts and Sciences, University of Pennsylvania. Accessed online January 27, 2017 https://lauder.wharton.upenn.edu/wp-content/uploads/2015/06/Chronology_Economic_Financial_Crisis.pdf

Hossain, N., L. Brito, F. Jahan, A. Joshi, C. Nyamu-Musembi, B. Patnaik, M. Sambo, A. Shankland, P. Scott-Villiers, D. Sinha, D. Kalita, and N. Benequista. 2014. "'Them Belly Full (But We Hungry)': Food Rights Struggles in Bangladesh, India, Kenya and Mozambique." Synthesis report from DFID-ESRC research project "Food Riots and Food Rights." Brighton: Institute of Development Studies.

International Monetary Fund (IMF). 2007. *World Economic Outlook: A Survey by the Staff of the International Monetary Fund*. Washington, DC: Occasional paper/International Monetary Fund.

Keating, J. 2013. "War of Ideas: What Can We Learn From the Last 200 Million Things That Happened in the World?" *Foreign Policy Magazine*. Accessed February 13 http://foreignpolicy.com/2013/04/10/what-can-we-learn-from-the-last-200-million-things-that-happened-in-the-world/

Kenyon, P. 2007. "Striking Textile Workers in Egypt Get Concessions." *National Public Radio Morning Edition*, October 12, 2007. Accessed February 10, 2017 www.npr.org/templates/story/story.php?storyId=15218389

Kotz, K. 2013. "Anonymous: Rehtaeh Parsons' Rapists Will Be Held Accountable." *Salon*. Accessed February 9, 2017 www.salon.com/2013/04/11/anonymous_rehtaeh_parsons_rapists_will_be_held_accountable/

Kumhof, M., and R. Rancière. 2011. "Leveraging Inequality." *International Journal of Labour Research* 3 (2): 189.

Lakner, C., and B. Milanovic. 2015. "Global Income Distribution: From the Fall of the BerlinWall to the Great Recession." *The World Bank Economic Review*, World Bank.

Leetaru, K. 2014. "Database of Events Language and Tone (GDELT)." Accessed online November 15, 2014 www.gdeltproject.org/

Li, Y. 2010. "Collateral Damage From the Global Financial Crisis: Could Developing Countries Land in Another Round of Debt Crises?" *Responding to Challenges Posed by the Global Economic Crisis to Debt and Development Finance*, UN Conference on Trade and Development (UNCTAD).

Lin, J., and V. Treichel. 2012. "The Unexpected Global Financial Crisis: Researching Its Root Cause." Policy Research Working Paper No. 5937. Washington, DC: World Bank.

May 12 Coalition. 2011. "Real Revenue Options to Restore the Worst Cuts." Accessed Online March 1, 2017 https://nocutsny.files.wordpress.com/2011/06/may-12-coalition-real-revenue-options-final-5-27-11.pdf

Milanovic, B. 2012. "Global Income Inequality by the Numbers: in History and Now – An Overview." Policy Research Working Paper No. 6259. Washington, DC: World Bank.

Milanovic, B. 2016. *Global Inequality: A New Approach for the Age of Globalization*. Cambridge, MA: Harvard University Press.

Moghadam, R., C. Cottarelli, and O. Blanchard. 2011. "Managing Global Growth Risks and Commodity Price Shocks – Vulnerabilities and Policy Challenges for Low-Income Countries." International Monetary Fund paper prepared by the Strategy, Policy, and Review Department, Fiscal Affairs Department, and Research Department, in collaboration with the African Department and the Asia and Pacific Department.

Mongelli, F. 2013. "The Mutating Euro Area Crisis: Is the Balance Between 'Sceptics' and 'Advocates' Shifting?", *Occasional Paper Series No. 144 / February 2013*. European Central Bank. Accessed online February 17, 2017 www.ecb.europa.eu/pub/pdf/scpops/ecbocp144.pdf

Mouhoud, E. M. 2012. "The Political Economy of Arab Revolutions: Analysis and Prospects for North-African Countries", *Mondes en développement* 2/2012 n°158:35–50. doi:10.3917/med.158.0035.

Musembi, C., and P. Scott-Villiers. 2014. "'The Constitution Lies to Us:' Securing Accountability for the Right to Food in Kenya." Food Riots and Food Rights project report. Brighton/Nairobi: Institute of Development Studies/University of Nairobi. www.foodriots.com

Ní Chasaide, N. 2013. "The Battle for Debt Justice." *International Policy Analysis Paper*, Eds. Puschra, W. and Burke, S. February 2013, Friedrich-Ebert-Stiftung, New York.

Obstfeld, M., and K. Rogoff. 2009. "Global Imbalances and the Financial Crisis: Products of Common Causes." SSRN Scholarly Paper ID 1533211. Rochester, NY: Social Science Research Network. https://papers.ssrn.com/abstract=1533211

Oikonomides, T. 2013. "The 'Squares' Movement: Combining Protest and Solidarity." *International Policy Analysis Paper*, Eds. Puschra, W. and Burke, S. February 2013, Friedrich-Ebert-Stiftung, New York.

The Organization for Economic Cooperation and Development (OECD). 2010. *OECD Economic Outlook No. 87*. Paris: OECD.

Ortiz, I., and M. Cummins. 2013. *The Age of Austerity: A Review of Public Expenditures and Adjustment Measures in 181 Countries*. New York and Geneva: Initiative for Policy Dialogue and the South Centre.

Ortiz, I., M. Cummins, J. Capaldo, and K. Karunanethy. 2015. "The Decade of Adjustment: A Review of Austerity Trends 2010–2020 in 187 Countries." *Extension of Social Security Series No. 53*. Geneva: International Labour Office.

Ortiz, I., S. Burke, M. Barrada, and H. Cortés. (2013). Working Paper: "World Protests 2006–2013." New York: Initiative for Policy Dialogue and Friedrich-Ebert- Stiftung, New York Office. http://policydialogue.org/files/publications/World_Protests_2006-2013-Complete_and_Final_4282014.pdf

Ostry, J. D., P. Loungani, and D. Furceri. 2016. "Neoliberalism: Oversold." *Finance & Development* 53 (2): 38–41.

Palley, T. 2007. "Financialization: What It Is and Why It Matters." *Political Economy Research Institute Working Paper Series*, Number 153, University of Massachusetts, Amherst.

Pisani-Ferry, J., and A. Sapir. 2009. "Banking Crisis Management in the EU: An Interim Assessment." *Bruegel Working Paper 2009/07*. Bruegel.org. Accessed February 17, 2017 http://bruegel.org/wp-content/uploads/imported/publications/wp_banking_lessons_091209.pdf

Romero, G. Z. 2013. The Future Is in the Making: A Year-and-a-Half After 15 May." *International Policy Analysis Paper*, Eds. Puschra, W. and Burke, S., February 2013, Friedrich-Ebert-Stiftung, New York.

Rude, C. 2005. "The Role of Financial Discipline in Imperial Strategy." *Socialist Register 2005: The Empire Reloaded* 41: 198–222.

Rude, C. (2010). "The World Economic Crisis and the Federal Reserve's Response to It: August 2007–December 2008." *Studies in Political Economy* 85 (1): 125–148.

Schneider, M. 2008. "We Are Hungry!" A Summary Report of Food Riots, Government Responses, and States of Democracy in 2008. Mimeo. Ithaca, NY: Cornell University.

Schrodt, P. 2012. CAMEO: *Conflict and Mediation Event Observations Event and Actor Codebook*. Pennsylvania State University. Accessed February 13, 2017 http://data.gdeltproject.org/documentation/CAMEO.Manual.1.1b3.pdf

Slifer, S. 2013. "Rehtaeh Parsons Update: Police Reopen Alleged Rape Case of Canadian Girl Who Committed Suicide." *CBS News*. Accessed February 9, 2017 www.cbsnews.com/news/rehtaeh-parsons-update-police-reopen-alleged-rape-case-of-canadian-girl-who-committed-suicide/

Smith, H. 2010. "Greek Protesters Storm the Acropolis." *The Guardian*. Accessed online February 17, 2017 www.theguardian.com/business/2010/may/04/greek-protesters-storm-acropolis

Stiglitz, J. E. 2002. *Globalization and Its Discontents*. New York: W.W. Norton.

Stiglitz, J. E. 2009. Report of the Commission of Experts of the President of the United Nations General Assembly on Reforms of the International Monetary and Financial System, September 21, 2009. Accessed online January 27, 2017 www.un.org/ga/econcrisissummit/docs/FinalReport_CoE.pdf

Stiglitz, J. E. 2014. "Lessons of the North Atlantic Crisis for Economic Theory and Policy." In *What Have We Learned? Macroeconomic Policy After the Crisis*, edited by G. Akerlof, O. Blanchard, D. Romer, and J. Stiglitz. London, UK: International Monetary Fund, MIT Press.

Tierney, B. 2011. "US Uncut." *Counterpunch Magazine*. Accessed online March 1, 2017 www.counterpunch.org/2011/04/18/us-uncut/

United Nations. 2011. *The Global Social Crisis: Report on the World Social Situation 2011*. Economic and Social Affairs. New York: United Nations Publications, 2011. Accessed January 27, 2017 http://social.un.org/index/linkclick.aspx?fileticket=v0lqqd2ft3k=&tabid=1561

Warren, A., and A. Dirksen. 2014. "Augmenting State Secrets: Obama's Information War" *Yale Journal of International Affairs* IX (1): 68–84, Winter.

Wedes, J. 2015. "Focus On: The #Right2Water Movement." *The Global Platform* blog, September 16, 2015. Accessed online March 1, 2017 http://blog.global-platform.org/2015/09/16/focus-on-the-right2water-movement/

Whelan, K. 2013. "Ireland's Economic Crisis: The Good, the Bad and the Ugly." University College Dublin Centre for Economic Research Working Paper Series; WP13/06. Accessed online February 17, 2017 www.ucd.ie/t4cms/WP13_06.pdf

Wikileaks. (2010a). "Afghan War Diary." *Wikileaks.com*. Accessed online January 27, 2017 https://wikileaks.org/afg/

Wikileaks. (2010b). "Iraq War Logs." *Wikileaks.com*. Accessed online January 27, 2017 https://wikileaks.org/irk/

Williamson, J. 2004. "The Washington Consensus as Policy Prescription for Development." A lecture in the series "Practitioners of Development" delivered at the World Bank on January 13, 2004. © Institute for International Economics. Accessed February 7, 2017 https://piie.com/publications/papers/williamson0204.pdf

World Bank. 2010. *World Development Report 2010: Development and Climate Change*. Washington, DC: World Bank. Accessed February 17, 2017 https://openknowledge.worldbank.org/handle/10986/4387 License: CC BY 3.0 IGO.

World Bank. 2011. "Responding to Global Food Price Volatility and Its Impact on Food Security." *Background paper for April 4, 2011 meeting of the Development Committee (Joint Ministerial Committee of the Boards of Governors of the Bank and the Fund on the Transfer of Real Resources to Developing Countries) prepared by the staff of the World Bank*. Accessed February 17, 2017 http://siteresources.worldbank.org/DEVCOMMINT/Documentation/22887406/DC2011-0002(E)FoodSecurity.pdf

Yousfi, H. 2013. "UGTT at the Heart of a Troubled Political Transition." *The Future We the People Need: Voices From New Social Movements in North Africa, Middle East, Europe & North America*. FES International Policy Analysis February 2013. Friedrich-Ebert-Stiftung, New York.

3 Framing 'food riots'

Subsistence protests in international and national media, 2007–2012

Naomi Hossain, Devangana Kalita, Bonface Omondi, Lucio Posse, Vaibhav Raaj, Muhammad Ashikur Rahman, and Michael Sambo

Introduction

This chapter investigates some of the conceptual and methodological challenges of making sense of contemporary 'food riots', particularly the wave of protests reported worldwide during the period of the global food price spikes, around 2007–2012. It does so through an exploration of the framing of media reportage of food-related protest events during this period. Media coverage frequently provides the empirical basis for cataloguing and studying 'food riots' and other forms of contentious political events, as in Chapter 2 in this volume, *A world in protest*. The aim in this chapter is to provide an analysis of how newspaper coverage framed food riots and related events during the global food crisis period, and some discussion of the effects of such coverage.

Framing 'food riots': Frames of collective action and the media

What do we mean by 'framing', and why might it matter to the analysis of food riots? There are two disciplinarily distinct applications of the concept of 'framing', and both are relevant here. The first is the idea of framing as 'meaning making' in relation to social movement analysis (Benford and Snow 2000). Frame analysis helps ensure that

> social movements are not viewed merely as carriers of extant ideas and meanings that grow automatically out of structural arrangements, unanticipated events, or existing ideologies . . . [but] as signifying agents actively engaged in the production and maintenance of meaning for constituents, antagonists, and bystanders or observers.
>
> (Benford and Snow 2000: 613)

Frames help explain collective action by drawing attention to the work that goes into making meaning, to produce "action-oriented sets of beliefs and meanings that inspire and legitimate the activities and campaigns of a social movement organization" (Benford and Snow 2000: 614). Collective action frames during food riots would, we expect, include a strong element of moral economy thinking, particularly demands for public action against food price rises, justified by a shared sense that the right to subsistence is prior to the right to profit from dearth (Thompson 1971; Thompson 1991).[1] Collective action frames of subsistence protests, particularly moral economy ideas, are discussed in detail elsewhere in this volume, in chapters exploring the grievances that motivate collective action, and how those grievances gain moral and political purchase through mobilization and interaction with political and policy elites. This chapter draws on those insights from other chapters here for some of the analysis.

Most of this chapter concentrates on addressing some preliminary questions about media framing of food riots. It draws on analysis of the frames in selected national and international newspapers, and discusses their effects. Media framing is a core analytical concept in the study of media effects (Entman 1993; Scheufele 1999; Matthes 2009). By framing as a media effect, we mean that we expect that newspaper coverage will shape how 'food riots' come to be constructed in media and thereby in public discourse more broadly; however, we also recognise that these effects will interact with the individual and group worldviews in order to acquire meaning: media frames are not simply given by news outlets, but constructed in collaboration with audiences. This social constructivist view of media framing (Scheufele 1999) encourages attention to how the media "select some aspects of a perceived reality and make them more salient in a communicating text, in such a way as to promote a particular problem definition, causal interpretation, moral evaluation, and/or treatment recommendation" (Entman 1993: 52). In the instance of food riots during 2007–2012, this means attending to where and which events are reported by whom, how actors are described and actions explained, and the kinds of resolution achieved.

As framing is not an effect achieved by the producers of media alone, it is also necessary to examine how consumers, and consumers' pre-existing ideas, shape frames. To look at how pre-existing audience narratives may construct 'food riots', we need to take into account the influence of ideological positions, particularly among global elites, and the tone of international coverage, particularly any efforts to appeal to 'human interest' or threats of global crisis. But we need, too, to consider the media as an active participant in the politics of provisions, whose framing is powerful in how it focuses public attention and articulates versions of public opinion. This chapter focuses chiefly on a discussion of media framing, looking at how different kinds of newspapers and other news sources with different audiences framed 'food riots' and related protests during the global food crisis period (mainly focusing on 2007–2012). This exercise has been undertaken in an effort to understand any systematic biases, gaps, or resonances between the frames of collective action and those shaping news coverage of food riots. This may further improve the scope for analysis using newspaper coverage, helping to avoid errors and strengthen the basis for uncovering the key mechanisms at work in food riots.

To address these issues of how food riots are framed, the chapter first explores how international media identified, framed, and reported on such events during the recent global food crisis. The chapter then discusses some of the issues arising in an effort to construct political event catalogues from national news sources in four countries. The chapter then goes on to contrast findings from the national events catalogues with those of in-depth primary research into key incidents and struggles reported in the media. This provides an opportunity to discuss the lessons and limitations of qualitative case study research into such contentious political events. The chapter concludes with a discussion of the areas of overlap and dissonance between collective action frames and media frames regarding food riots. It also discusses advantages and limitations of mixed methodological approaches to the study of food riots and related modes of contentious food politics.

International media coverage of 'food riots', 2007–2012

The politics of provisions are mainly negotiated at the national level (Bohstedt 2010), but participation in a globalized food economy introduces supra-national dimensions to struggles over food (Patel 2012; Holt-Giménez and Patel 2012). In 2008, the food price crisis hit food systems all around the world, and the aggregated response of food riots in as many as 37 countries was seen as a wave of great significance; these events were cited collectively as evidence of a problem in the global food system, affecting all. Protectionist responses, particularly in large food producers, further highlighted the interdependence of national food systems with the global food economy. Climate change-related events led to a second global staple price spike in 2010–2011, further confirming the international and interlinked nature of the crisis.

The apparently global and contagious nature of the food price crisis and of the transnational and national policy responses it engendered means it is important to situate individual country experiences with 'food riots' in a more global perspective. This suggests it may be important to understand how food riots and/or the global food crisis are framed as a distinctly global problem, as distinct from a problem in poor countries far from the centers of global power. However, as we still expect that the most direct mechanisms of effect are likely to be policy levers at the national level, we still need to understand framing devices and strategies at that level.

Approach to exploring food riots in the international media

To study the coverage of food riots in the international media we used the online search engine of Nexis® (nexis.com), the international news database. Nexis allows searchable access to 609 news sources, including aggregators and newswire services, as well as online resources such as blogs and magazines. Although we used a variety of search terms and approaches in our search strategy, we focus here on the results from our search for and analysis of news coverage of incidents labelled 'food riots' during the period 2006–2016. Our search was worldwide, and included all available news sources in the Nexis database. Our Nexis searches identified 6,301 news items that included the term 'food riot' in the period 2006–2016.[2]

Figure 3.1 offers a sense of the timeline of media coverage of food riots. It shows that by far the greatest number of articles including mention of the term 'food riot' anywhere in the text during the period 2006–2016 were published in 2008, when 2,545 news items were identified. As Figure 3.2 of the global food price index of the Food and Agriculture Organization (FAO) during the same period shows, 2008 was the year of the first major price spike. However, whereas global food prices spiked again, peaking even higher in real terms in early 2011, international news attention to food riots events fell to 1,073 stories, only 42 per cent of the number of stories in 2008.

In order to explore the representation of food riots in a closer and more structured manner, we analyzed 156 articles from the peak period of the global food crisis (2007–2012) in which 'food riot' featured in the headline, and therefore could be assumed to be a core theme. This 156 represents more or less the full set of articles, other than a small number removed because the term referred to events very far removed from the subject of dearth and popular protest (for instance, teenaged vandals in an Edinburgh burger chain)[3] or because of re-publication of near-identical article. A simple coding was applied to the articles using NVivo10, in order to enable frame and textual analysis of the set. After an initial scan, one set of codes classified articles according to the general article type, specifically whether they were: a) factual accounts of protests, b) opinion pieces or analyses of the food crisis, or c) reporting on or providing a platform for advocacy, promoting particular responses to the food crisis. Just under half, or 76, of the articles were coded as containing factual accounts or news stories of protests, and only

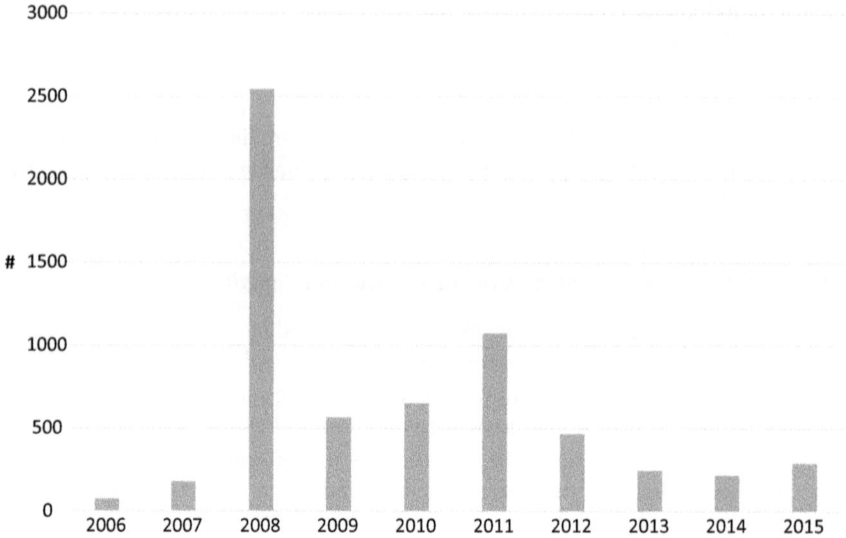

Figure 3.1 Number of news reports mentioning 'food riot', 2006–2015

Source: Authors' calculations from LexisNexis searches, 24–26 January 2017

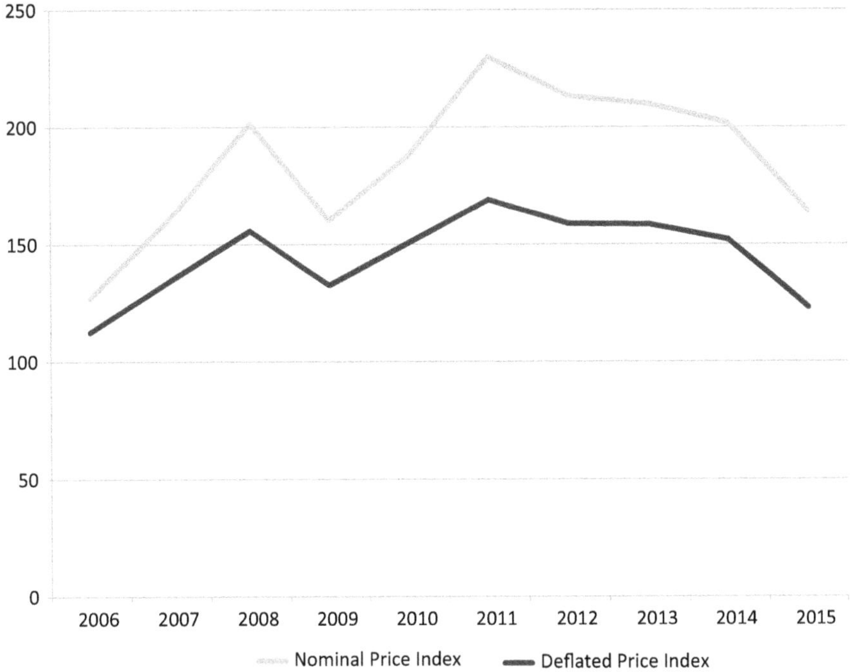

Figure 3.2 FAO Food Price Index, 2006–2015

Source: FAO, www.FAO.org/worldfoodsituation/foodpricesindex/en –accessed 31 January 2017

four of these also contained significant analysis or advocacy content; 59 articles contained opinion or analysis of the food crisis; 37 were advocating or giving a platform to advocacy for a policy response or course of action; 16 did both.

Overall, then, while around half of the articles in the international press which headlined 'food riots' during this period were accounts of actual protests, more offered opinion or advocacy. It should be noted that the Nexis database predominantly comprises international news sources, and by no means includes even all English-newspapers from developing countries – and so that first response articles, when the events were new or unfolding, may have been covered in local and vernacular media without ever making it into the international media as news stories in their own right. In reporting on a major global crisis such as the food price spikes of 2007–2012, then, the international media may of necessity be playing the role of an aggregator and commentator more than a news source.

In the absence of any substantial analysis of the framing of food riots in the international media to our knowledge to date, we focus here on addressing some preliminary questions: what are some of the key frames within which food riots are reported? What are the messages they send, in terms of the information they communicate about such protests and the responses they suggest or call for? Taking into account the elements of Entman's definition of framing, we turn now to an examination of: a) the 'problem definition', b) the 'causal interpretation', c) the 'moral evaluation', and d) the 'treatment recommendation' that emerges from this body of material.

The problem with food riots

The broad definition of the problem within which the 'food riots' articles were framed was the global food crisis, generally associated with the rapid rise of staple grain prices on global and national markets in the period 2007–2008, and again in 2010–2011, and the consequent global subsistence crisis. However, as Figure 3.3

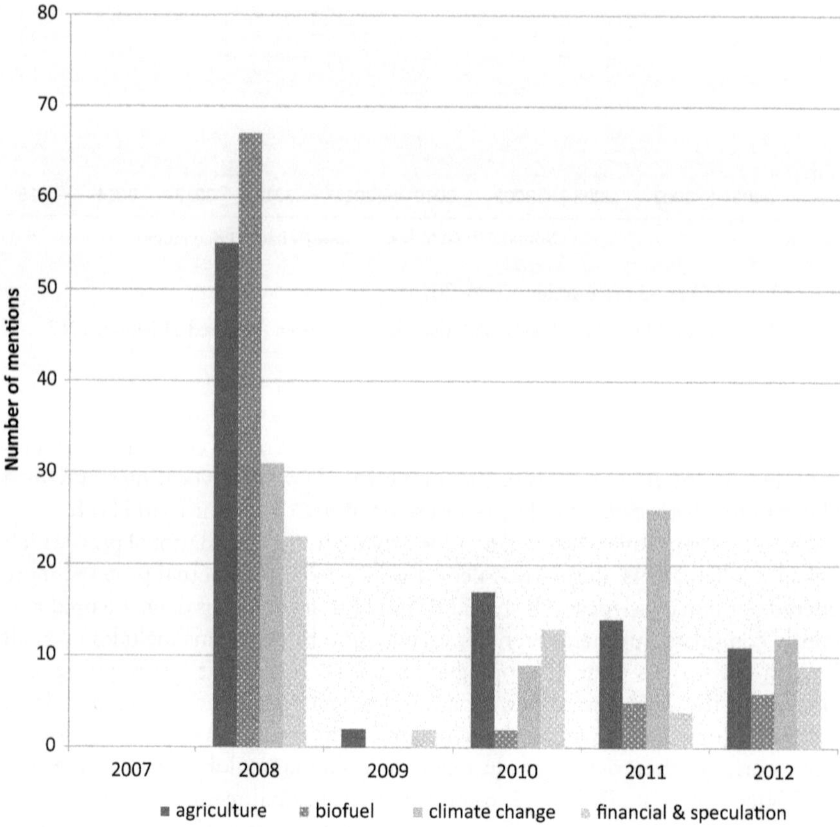

Figure 3.3 News article content on the causes of the global food crisis

Note: frequencies are of numbers of times causes are mentioned, rather than numbers of articles in which they are mentioned

Source: Authors' analysis of sampled articles from nexis.com

suggests, the content of news articles regarding the causes of the global food crisis about which people were protesting changed with changes in the nature of the crisis itself: whereas in 2008 a substantial part of the debate focused on bio fuels and financial speculation, by 2011, the focus was more firmly on climate change, as well as, to a lesser degree on agriculture.

The intensity and scale of food-related protest may have been as high as in 2008, and the protests had substantially shifted from low- to middle-income, chiefly Middle Eastern-region countries, but there were far fewer articles headlining 'food riots' in this second phase of the crisis. Why was there substantially less coverage of food riots during the second price spike than in 2008? There may be several reasons. The first may have been the absence of novelty, as food riots were no longer 'news' after the wave of protests in 2008. Second, global policy efforts to address both the food insecurity of vulnerable populations and the wider issue of price volatility were in train by this time, and the 2010–2011 price spikes may have been seen as more straightforwardly the result of the natural disasters associated with climate change. In other words, the facts of this second round of food riots had changed, so they became less newsworthy.

Other reasons for this difference between 2008 and 2010–2011 may include differences in context and of surrounding events. More of the protests in the later period were now taking place in middle-income countries, including many in the conflict-ridden and predominantly authoritarian Middle East. These are more highly developed countries than the low-income countries that dominated food riots in the first phase, so that the equation of food riots with 'hunger' may have been less obvious. The narrative of food riots in Middle Eastern countries may also have been more complex because the international media may have been both more familiar with these settings than with the less developed countries of the earlier wave of food riots. Journalists may therefore have expected more complex political realities, so that the simple message sent with 'food riots' fit such protests less well. A complicating factor was also that the world, and in particular, the developed world on which the international media chiefly reports, was by 2010–2011 deep in the throes of the worst economic crisis it had seen in almost 80 years. While the second round of the food crisis was dramatic, there was no shortage of other news stories, including many of popular political protest around basic needs in the developed countries on which to report during this period. It is possible, then, that the 2010–2011 food crisis was, in part, drowned out by the global financial crisis.

There was also a narrower definition of 'the problem' of food riots, however, related to the problematic impacts and implications of mass popular subsistence protests. Many opinion pieces and advocacy articles conveyed a sense of the 'threat of food riots'. For instance, 'Food riots "an apocalyptic warning"' was the headline of one Australian Broadcasting Corporation Premium News story in 2008 (Barlow 2008). Another was the prediction that the world faced "an inflationary time bomb as shortages of food threaten to push prices to fresh all-time highs", by The Independent newspaper in August 2010 (O'Grady 2010). Of the substantial number of articles that were opinion or advocacy pieces, many warned that food riots would result should certain events occur (further price rises, natural disasters, failed harvests) or actions fail to be taken (price stabilization or consumer insulation in the short-term, agricultural investment or reform in the medium or long). A total of 82 of the 156 articles were coded as containing a prominent warning or reference to the 'threat' or 'fear' of food riots, most of which (66) were also classified as either opinion or advocacy pieces (factual articles reporting on food riots were less likely to include such warnings, with only 22 doing so).

But why should food riots pose a threat, and to whom or what? This is often left implicit, as if subsistence-related protest is an obviously bad thing to be avoided at any cost. Four specific threats can be discerned. First, at the local level, there may be violence and a threat to human life. Our calculations from the set of 156 articles indicate that at least 100 people were reported as killed in food riots during 2007–2012, and at least 2,000, including police, army, and peace-keeping force members, were injured – more than 800 in Algeria alone. Unruly politics are in general a threat to the public authorities, by their very definition (Khanna 2012).

Second, and again at the local level, large protests can disrupt economic and social life. Where food traders or markets are specifically targeted, this may even have counter-productive effects in the short-term, as fearful traders retreat or conceal their wares. During the Mogadishu food riots of May 2008, news reports noted that the 20,000 reported protestors explicitly targeted food traders who demanded payment in U.S. dollars instead of the (often counterfeited) Somali shillings:

> "We do not need cruel traders", "We do not need people who are in love with dollar yet we have the shillings", "We want our right to live", they chanted in the streets, where plumes of smoke rose.
>
> "We are going to demonstrate until they (traders) accept the Somali shilling . . . we cannot afford dollars," one protestor Hassan Said told AFP.
>
> "And if they refuse to lower food prices, we will start looting their shops so that we can survive."
>
> (Abdinur 2008)

Although little information about the identities of the protestors is provided, this Agence France Press news service account of their grievances (cruelty, corruption, official failure to curb rampant inflation), ideological justification for protest (the right to live, the struggle against inequality and injustice), and protest repertoires (bargaining over lower food prices or looting) describes a food riot very much in the European 17th and 18th century tradition documented by Charles Tilly (1974) and John Bohstedt (2010). As those scholars have shown, the food riot typically 'worked' at the local level, if only to temporarily enforce a local moral standard onto food markets and some momentary redistribution. For the owners of local businesses and the public authorities tasked with protecting their properties and livelihoods, such an episode is disruptive and costly, as well as dangerous.

A third threat food riots are framed as posing is to political elites. At the extremes, food riots are cast as 'a threat to democracy', as in The Associated Press article about the April 2008 Haiti food riots(Katz 2008). The then-President Rene Preval instructed food rioters to stop looting because the solution was "not to go around destroying stores", blaming the country's dependence on imported rice for the food crisis. His intervention came at a crucial moment, although (as subsequent events suggest), lacked the desired effect:

> The speech had been widely anticipated by observers who said Preval's response could determine the course of the demonstrations and of his government.
>
> "I believe if President Preval talks to the people about the high cost of living, people will listen to the president and go home," said Sen. Joseph Lambert, a former senate president and a member of Preval's party. "If not, if there is an attempt at a coup d'etat to remove the president, things will get worse."
>
> Food prices, which have risen 40 percent on average globally since mid-2007, are causing unrest around the world. But they pose a particular threat to democracy in Haiti, where most people live on less than $2US (€1.27) a day.
>
> *(Katz 2008)*

A short straight line is drawn here between the very low incomes of the Haitian people, the steep rises of staple food costs worldwide, and the risk to 'democracy', apparently represented by the U.S.-backed Preval regime. Food-related protest after Pakistan's worst ever floods in 2010 similarly rise to fears that in the worst-case scenario, "the military might feel compelled to step in if protests got out of hand" (Reuters 2010). Other articles also connect food riots and the interests of political elites, but do so without going quite so far as to suggest they jeopardize democracy itself. Instead, many articles associated the price crisis with loss of popular legitimacy by political leaders. For instance, Bolivia's Evo Morales was reported to have cancelled a public appearance at the height of the 2011 crisis, fearing popular backlash engineered by the opposition (UPI 2011).

Causes of food riots

It would appear that public authorities unable to guarantee basic subsistence are also unable to assure law and order. Arguably (although the argument is rarely made in the news articles reviewed here), a lack of official legitimacy manifests in generalized social unrest and a descent into lawlessness. But how does this come about? What is the step from dearth to delegitimated authority, and from there to anarchy? This is less clear, or more implicit. News media commonly use action terms to explain the causal relationship between price rises and social unrest such as- 'provoke', 'spur', 'trigger', or 'spread'.

These terms help frame food riots as a simple mechanism of societal response to dearth. In some instances, rioters' anger is due, in a simple way, to shortages or unaffordability. In others, such as the Mogadishu example given previously, the anger is linked to a perception of injustice that while they (the protestors) go hungry, others are immorally turning a profit. But a 'food riot' can be easily discerned as any protest by people who lack means in times of dearth, about which it is safe to assume the 'trigger' is the anger provoked by hunger and injustice.

The idea that food riots are somehow 'natural' is a particular political framing common to many on the left. Of the riots that hit three major Burkina Faso towns in February 2008, one commentator explained that:

> "This reaction was expected," Laurent Ouèdraogo, secretary general of the Confèdèration Nationale des Travailleurs du Burkina (CNTB) told IRIN, stating that the riots happened after anger welled up because of constantly rising prices for basics like food, cloth and petrol.
>
> "Misery does not wait and you see people witnessing everyday rising prices and they do not know what to do. The situation is like having matches near cotton that can catch fire at any moment," Ouèdraogo said.
>
> (IRIN 2008)

Like other trades union or leftist party spokespersons, Ouèdraogo is well-versed in the language of the popular threat from the discontented masses, and skilled at its use in advocating policy change – in this case, price controls. Similar points were made by trades unionists in the Philippines and leftist party spokespeople elsewhere.

But the automaticity of the threat from the discontented masses also appeared in the framing by or on behalf of groups strongly opposed to a leftist platform, such as the multilateral institutions, and specifically the economic globalization agenda of the International Monetary Fund and the World Bank. One finer-grained analysis contrasted the views on the North American Free Trade Agreement (NAFTA) of a former Mexican maize farmer who had marched in the 'tortilla riots' of 2007 against Robert Zoellick, the World Bank president who argued that far from raising the price of corn, free trade agreements kept it low (Koop 2008). But free trade and other open market policies were endangered by popular anger. The 'apocalyptic' ABC Premium News article cited previously reported the then-head of the International Monetary Fund, Dominique Strauss-Kahn, spelling out the reasons to be fearful of the effects of food riots, particularly in Africa:

> "Hundreds of thousands of people will be starving," he said.
>
> "Disruption may occur in the economic environment so that at the end of the day most governments, having done well for the last five or 10 years, will see what they have done totally destroyed and their legitimacy facing the population destroyed also."
>
> (Barlow 2008)

In other words, food riots jeopardized the entire neoliberal development project.

Moral judgements

Moral judgements of protestors were muted, or implicitly sympathetic, in the international literature. As we will see in the next section, national media coverage tended to be less universally sympathetic; closer proximity to events, local political interests and differences, and class or other social biases against particular protest groups within the media may contribute to this more heterogenous view of food-related protest. References to angry protestors drew attention to a simple mechanistic framing of the causes of food riots. By contrast, the many references to hunger and the ratcheting up of the subsistence struggles of people already on low incomes suggested a broadly sympathetic, or perhaps empathetic, view of food riots. This 'human interest' angle may also have been necessary in order for international news outlets to turn an abstract statistical account of global trade and food insecurity into a news story with wider appeal. But it has the overall effect of framing food rioters as driven to desperation – invoking a sense of a Hobbesian natural law of survival with universal resonance. An example of this is the use of descriptors of protestors: whereas almost half of all the articles described angry and violent situations using such terms, only 16 referred to protestors as 'angry', compared to the 42 in which they were described as 'hungry'. Notably, the factual or news story accounts were more likely to use the descriptor 'angry' and articles featuring more analysis and advocacy preferred 'hungry'.

The 'treatment recommendation'

The 'treatment recommendation', as Entman's (1993) definition of a frame has it, varies across the articles. One factor shaping that appears to be location and role within the global food system; for instance, there seemed to be a comparatively large number of Australian, Canadian, and Irish news articles among the set, perhaps reflecting those countries' relatively prominent roles as food aid providers. The phase of the food crisis being reported on also shaped the diagnosis of the problem and the proposed solution, as noted previously. Ideological factors may have shaped the 'take' on issues of free trade and the extent to which it was a problem or a solution, but most of the more detailed analyses examined arguments pro and con. A tentative conclusion may be that the global food crisis and the threatening spate of food riots that followed in general weakened the case for the financialization of food commodity trade (the business and trade press no doubt have a different framing). There appeared to be a relatively strong pro-intervention framing on humanitarian grounds, but not to the extent of national protectionist policies. Whereas many newspapers published statements and reports from the UN, IMF, and World Bank on the food crisis, the analysis and advocacy of groups such as farmers' and labor unions or consumers' associations rarely achieved great prominence in the international media, and appear to be more usually relegated to the statement of key grievances at the moment of, or in predicting, the food riot.

Media framing of subsistence protests in Bangladesh, India, Kenya, and Mozambique

Methodology and approach

This section of the chapter reports on experiences developing and using national events catalogues to support analysis of the causes and effects of popular mobilization around food in Bangladesh, India, Kenya, and Mozambique during 2007–2012 (Hossain *et al.* 2014). Developing these catalogues proved methodologically challenging and labor-intensive compared to the searchable online database used to analyze international media coverage. Nevertheless, the exercise generated some valuable insight into how the national media were framing subsistence protests – and to what effect – in each country.

Note that we explored media coverage of all kinds of subsistence protests relating to food during the period 2007–2012, rather than restricting our searches to 'food riots'. This is partly due to an early finding of a tendency for 'food riot' to be applied to explain events abroad, rather than at home, where grievances tend to be explained contextually and as more complicated matters. A second reason we focused on all manner of mobilization around food rather than just 'food riots' is in order to assess mobilization around the right to food, particularly in India, and other modes of organizing as civil society, as distinct from the more 'transgressive' forms of contention expressed in riots (McAdam, Tarrow, and Tilly 2001).

The *Food Riots and Food Rights* project selected Bangladesh, India, Kenya, and Mozambique – four food insecure countries with different experiences of popular political mobilization during the global food crisis – for closer and comparative analysis. Bangladesh was among the countries the international press and multilateral agencies listed as having experienced food riots in 2008; with a shared history of famine and famine response, India's active, broad-based Right to Food struggle had yielded the establishment of a Right to Food Act; Kenya had similarly instituted a new constitution in 2010, enshrining the Human Right to Food after the election violence of 2008; and Mozambique was among the countries in which food – and fuel – riots were prominently covered in the international media, including in 2010. The aim was to use national coverage of popular struggles relating to food in this period to understand the political effects of different modes of mobilization against the backdrop of a globalized food crisis.

A common approach was developed for the identification and selection of news sources, and for the analysis of individual stories and connections between them. This was designed to highlight the identities of protestors, their stated grievances and demands, the objects of their demands, repertoires or techniques of protest, and policy and political responses by ruling elites. The 'political event catalogues' so developed aimed to provide a sense of the universe of food-related protests reported, and that therefore had a chance of coming to the attention of political and policy elites, and to provide the basis for sampling cases to examine in closer detail through qualitative interviews with protest participants and with policy and political elites. The first challenge was the selection of credible national news sources. We concentrated on newspapers, as most large data studies rely on print media, and increasingly in particular on their digital archives. Sources were selected based on circulation, nature of the readership, ideological leaning, reputation, and the availability of online archives and search engines.

In Bangladesh, the online archives of *Prothom Alo* and *The Daily Star* were used. Both are part of the same news organization and have liberal, progressive inclinations, and both emerged with a new wave of print media that emerged with Bangladesh's return to democracy in the 1990s. *Prothom Alo* has the highest circulation of vernacular newspapers and *The Daily Star* leads the English daily papers. The more ideologically and socially conservative *Dainik (Daily) Ittefaq* has been in circulation since the 1950s, and remains an authoritative source, particularly in rural areas. The *Ittefaq* had no online archive at the time, and 12 university students were recruited to hand-search seven years of the physical archives of the newspaper in the library of the University of Dhaka.

In India, the large number of national, regional, and local newspaper meant no individual newspaper was likely to cover popular mobilizations across the expanse of this large and complex country. Two national dailies were finally selected for India: *The Times of India* and *The Hindu*, both established newspapers on opposite ends of the ideological spectrum. *The Times of India* is perceived as populist but also as pro-business and establishment, whereas *The Hindu* represents a largely leftist parliamentary perspective. *The Times* has the highest English-language circulation in the country. Both newspapers had searchable digital archives.

In Kenya, the search moved beyond print media as online searches of the two leading dailies, *The Nation Newspaper* and *The Standard Newspaper* yielded too few results to gain a sense of popular mobilization around food. Wider online searches yielded coverage in YouTube channels of Kenya Broadcasting Corporation (KBC), Kenya Television Network (KTN), and Citizen TV, which were then analyzed for their contribution to national coverage.

In Mozambique, digital archives were limited and the physical archives were unwieldy and challenging to research. These challenges meant that the news content searches focused on hand searches of selected media from the months surrounding the riots of February 2008 and September 2010. The Mozambique catalogue brings together a larger variety of sources from state-controlled to 'independent' (to whatever relative extent possible under the authoritarian regime) media, bringing together around 163 articles drawn from 15 different sources: three daily printed papers (*Notícias, Diário de Moçambique and O País*), six electronic dailies (*CanalMoz, Diário de Notícias, Ponto Certo, Vertical, Mediafax and Diário do País*) and six printed weeklies (*Canal de Moçambique, Savana, Zambeze, Magazine Independente, A Verdade and Domingo)*.

Searching for relevant political events

Having identified sources, a second challenge was to locate reports of relevant political events for inclusion in the catalogue. A set of keywords was devised for the first round of basic search in all four countries. However, in the case of online archives, what these terms yielded depended on the quality of the search engine. For instance, the search term 'food riot' produced 573 entries in *The Daily Star* search engine. After scanning more than one-third of these articles, only one entry was found to be relevant, referring to food-related protests within Bangladesh. Adapting the search strategy after this first round, a second round of search terms was adapted with specific reference to the events known to have taken place in each country. For instance, in Bangladesh phrases such as "RMG (ready made garment) workers protesting for food hike" were tried, and linked terms such as *chapa* (minibus), *subsídios* (subsidies), *cesta básica* (basic basket), *revolta popular* (popular upheaval), and *custo de vida* (cost of living) were used for Mozambique. Once events were located, a closer search of the days before and after the event was undertaken to identify follow-up stories and precursors.

In building the national catalogues, the definition of a 'political event' was broad. The entries in the catalogues cover a wide spectrum of mobilization, from street demonstrations, rallies, riots and other forms of direct action, to public meetings, seminars, and petitions. This approach to the identification of events as part of wider struggles was motivated by Tilly's (2008) argument that to explain contentious struggle, it is not enough to count and catalogue individual events by type and draw correlations; an understanding of contentious politics depends on being able to make sense of individual events set within and as part of wider and more enduring struggles. This will entail not only observing the actions (the event counts), but also making sense of their effects and the responses they stimulate, and in addition, how the success or failure or repression of protest then feeds back into shaping how and when further protests take place. The aim of such analysis is to establish a mechanism-based approach, able to 'specify what sort of event produces the correspondence between the presumed cause and the presumed effect' (McAdam, Tarrow, and Tilly 2008: 309).

The catalogue entries included context, actors involved, grievances/demands made and to whom, the form of action, and the nature of response, including repressive or violence by state-sponsored actors. Information was also sought on the words used to refer to or represent the event and its actors, and statements recorded from event participants and those holding power. However, few media reports were exhaustive enough to provide the information sought, and there were few cases in which we were able to build a detailed narrative of a struggle across all its phases and interactions.

However, the events catalogues provided glimpses into the politics of provisions, with preliminary insights into who was protesting about what and how in each country. However, even extensive hand-searches of events known to have occurred failed to yield the detail necessary to gain a sense of why such groups came to mobilize, how they articulated their grievances and selected their repertoires for action, or of how the political and policy elites responded, and why. Critically, we identified significant instances of food riots and more enduring subsistence struggles that had gone largely unrecorded by the national media, notably in Kenya. In Madhya Pradesh, although the media regularly reported on starvation deaths and malnutrition, local activism around the right to food was substantially excluded from media coverage. In Bangladesh, events that were labelled 'food riots' in the international press were framed as significantly more complex matters in the national press, which situated them within their longer history as wage struggles.

Media bias

An urban bias, particularly in favor of coverage of events affecting capital cities, was common in national newspapers. In a small and selective way, we were able to confirm this bias, identifying events that went unreported. For instance, one of the Indian case study sites, in Madhya Pradesh, witnessed a series of protests and actions by a local organization which neither selected newspaper covered. Deaths from malnutrition in this state had focused media attention on issues of food security in Madhya Pradesh, but this interest did not apparently extend to coverage of popular mobilization relating to hunger. The coverage by *The Hindu* chiefly focused on events in Delhi, Karnataka, Kerala, Andhra Pradesh, and Tamil Nadu. In Kenya, similarly, local protests relating to claims to food relief that were identified during field work in Mathare and Ikutha had received little or no media attention. In Mathare, frequent skirmishes at the depots where the urban poor tried to buy subsidized maize meal were never reported. In Ikutha, three big protests against malpractice in the local distribution of relief food went unreported.

Events involving violence or disruption by economically or electorally important participant groups had greater coverage and space, often on the front page, and in many cases reporting framed such protestors as trouble-makers, stirring up the crowd. Incidents of state repression and violence were, by contrast, downplayed or given less attention, with the exception of Mozambique. This attention to state repression in Mozambique was because of the novelty of the underclass mobilizing and protesting the food and fuel price hikes, a protest widely considered legitimate although repressed by the state police. At the same time, the police were thoroughly resisted by the protesters who wounded some police officers and burned some police vehicles. Moreover, some private press may have wanted to highlight the intolerance of the Mozambican state, as the right to protest is protected in the national constitution, and state efforts to impede protest resulted in violence. There was a formulaic nature to these accounts, with a narrow range of descriptions and limited direct reporting of protestors' viewpoints. Histories and backgrounds of protest campaigns, movements, or eruptions were generally ignored.

The prevailing political climate of the country also determined the news which was produced and in what way. The extent of media coverage the 'riots' in Maputo received in 2008 and 2010 was higher than in 2012, when the massive presence of riot police at the main focus of riot incidents aborted the demonstrations. The authoritarian nature of the government in Mozambique also imposes various limitations to what things can be reported and how, limiting the scope for sympathetic coverage in fear of state reprisal. A focus group discussion with Bangladeshi journalists uncovered a tendency to exaggerate food-related gatherings as 'protests' during the military-backed caretaker government, so that "a queue of people waiting to purchase subsidized rice from the OMS (Open Market Sales programme) could get reported as an agitation for essential goods", as one journalist put it. This indicated a perception on the part of the media of their own role in holding power to account, and its particular importance during periods of undemocratic rule.

The construction of a 'food riot'

Comparatively few references to the term 'food riots' were present in the national media reports, and with the exception of Mozambique, 'food riots' invariably referred to events that occurred elsewhere; conversely, events labeled as 'food riots' by the international media were rarely labeled as such in the national media. For instance, while a search for 'food riot' in Bangladesh leaded only one relevant result in the national English newspaper, there were reports of 'food riots' in sub-Saharan Africa, India, and elsewhere. Similarly, the Indian media labeled garment workers' protests in Bangladesh as 'food riots', while the Bangladeshi media framed them as wage struggles. *The Hindu* tends to cover West Bengal fairly closely, yet carried only one piece reporting on the widespread and significant events the international media had marked as 'food riots' in 2007. The sole article on these protests reported the state government's claim that the riots were politically motivated and thus inauthentic. It should be noted that the government of West Bengal was at that time under the rule of the Communist Party of India (CPI-M), with whom *The Hindu* is ideologically aligned and which was facing a crisis of legitimacy after 34 years of rule.

The reluctance shown by the national media to frame incidents of domestic unrest related to food as 'food riots' suggests that the effect of such reports is a pointed political accusation: that the government of the country in question has failed in basic functions of governance, and is vulnerable to loss of authority and legitimacy.

National coverage tended to avoid framing subsistence protests as simple correlations between hunger, anger and protest. Instead they were often linked to other contextual factors such as party political competition; histories of struggle and enduring grievances among protesting groups; local food market conditions; the performance of local government and agencies; and the role of external actors in 'stirring up trouble'. Despite their frequent lack of sympathy with protestors and their means of protests, these articles aim to hold state power to account. See this excerpt from a December 2009 editorial in *The Daily Star* of Bangladesh, headlined: "*Why food price hike worries us*":

> Last but not the least, price hike of food and other essential commodities may lead to serious social unrest. It creates discontent, violence and food riots. We saw how several thousand garment workers at Fatullah, Narayanganj clashed with the law enforcers and smashed cars and vandalised factories on April 12, 2008 in anger at price hike of food and other essentials and

> low wages. In 2008, food price hikes sparked food riots in some African and Asian countries including Ethiopia, Cameroon, Haiti and the Philippines. In view of the fact that food is the most important of all the basic necessities in human life and per capita income in Bangladesh is one of the lowest in the world, and that it is the constitutional responsibility of the state to make provision for the basic necessities of life, including food, no government can avoid the responsibility of making food available to the people at affordable prices.[4]

While the article draws on the 'threat of food riots' trope of the international media, it avoids labelling garment workers protests in such terms, framing them instead as an early warning of the need for urgent state action, to prevent escalation into food riots as had occurred elsewhere. Another entry in the catalogue reports on a statement by the chairman of the Regulatory Reforms Commission makes the Amartya Sen argument that "famines takes place in the countries run by unelected governments at there is none to put pressure on the government to take the right decision".[5] This strong emphasis on the state responsibility and intervention. This is in strong contrast to 'treatment recommendation' framings in the international media, which concentrated around issues such as regulation of the financialization of food commodity trade and around investments in agriculture.

The widespread protests against local PDS stores and dealers in West Bengal were specified as 'ration riots' by the Indian national press and as 'food riots' by the international media. A Reuters report claimed, "the riots have been a major embarrassment for the state's ruling communists since it involves people living below the poverty line, the vote bank of the left for the last 30 years".[6] It located the cause of the riots in the problem of corruption, in line with the growing public sentiment that "for years, they have been cheating us".

Three entries deploy the term 'food riot' in relation to the West Bengal PDS riots. A *Times of India* piece in 2007 reports on the Congress party 'hitting back' at the CPI-M for "raging food riots across the state", with the Congress spokesperson terming it as an "anachronism . . . in a country which is proud to flaunt a booming economy". This 'attack' by the Congress is analyzed in the article as 'retaliation' against the Left's refusal to submit to the then-Congress-led coalition government over the unrelated issue of a U.S. nuclear deal agreement. An article in *The Hindu* the following day reported a rebuttal from the CPI-M politbureau, stating, "there are no food riots in West Bengal". It accused the opposition of "inciting violence on the issue of ration distribution", and blamed ration cuts by the central government. However, the retrospective use of 'food riots' in a *Times of India* article two months later reported that:

> [T]he recent food riots in West Bengal were the result of a corrupt and privatized public distribution system running with political patronage of the Left parties, a survey conducted by SC-appointed special commissioners on food security said. The scathing report has debunked the West Bengal

> government's stand that the riots were a reaction to decreased allocation of wheat and rice for the above poverty line quota. The commissioners have concluded that this could not have been the provocation for people running riot in rural West Bengal against lack of food grain.[7]

This suggests tensions within the national media over the construction of events as 'food riots', particularly in relation to the implications for state legitimacy and competitive party politics. A more limited review of regional media in two prime areas of the protests adds another layer to the narrative. An important part of the context which both the international and national media miss is that these 'ration riots' took place in the aftermath of a strong challenge to the hegemony of the Left Front regime in West Bengal. This took the form of massive resistance to official land acquisition efforts on behalf of Tata Motors. The West Bengal government responded with brutal police action, and finally withdrew. The Nandigram Singur resistance consolidated general disenchantment with the Left Front government, providing a political opportunity that played a role in the eruption of the PDS riots. The West Bengal media reported extensively on the riots, but neither malnutrition, nor the Right to Food, nor even the structural problems plaguing the PDS, nor the immediate grievances of local beneficiaries were central to its coverage. Instead, the reportage focused on events as they unfolded and the political analysis on the role of local units of the CPM. The state-level activists and experts we interviewed told us that political commentary in the Bengali press placed the PDS riots within along the continuum of protests against land acquisition and the call for '*portiborton*', or change, especially since the location of riots was close to Nandigram, the site of the resistance to the land acquisition program.

The media as an actor in the contentious politics of provisions

The media often plays a crucial role in the contentious politics of provisions, and can be seen as a key actor in framing and amplifying popular grievances and interactions between protestors and policy and political elites. In Kenya, even though the local chapters of *Bunge la Mwananchi* (the Ordinary People's Parliament) had been hard at work mobilizing around the rising cost of living since 2005, it was only when they intertwined the issue with the issue of suppression of press freedom that the media paid attention. The media coverage given to their high-profile infiltration and disruption of the Jamhuri (Independence) Day celebrations in December 2008, when several *Bunge* members and journalists were arrested and assaulted by the Presidential Guard, marked a crucial point in the emergence of the movement. The protests that have now come to be identified as the "Unga Revolution" first featured in public discourse when the national media reported the heckling of the minister for labour – who was reading the president's speech – by a diverse gathering of workers at the official Labour Day celebrations on 1 May 2009. After these successful episodes of gaining the media stage, grabbing state spaces as a form of mobilization and to express grievances became closely connected with the media attention such direct action would generate.

The role of the media also emerged as important in the context of weak functional channels of communication between protestors and government. In Mozambique, protests had no evident leaders, and the urban poor – the population most affected by the food price rises and subsidy cuts – had no political or civil society organizations to represent their interests to the authoritarian regime. In this communication vacuum, the period preceding the riots saw extensive media coverage of the rising prices of goods and services, alerting the government to the threat of popular mobilization due to price rises, and the media came to play a significant role in how events unfolded as protestors responded to reports by the relatively sympathetic independent media. Whereas ruling party-backed newspapers depicted the riots as illegal, opportunistic, and illegitimate due to violence and vandalism, the independent press frontally criticized the government and the police for the violently repressive response, which included live ammunition. The protests created a new space of articulation for academics, political analysts, and some politicians, who used the mass media to express dissent. The role of the media as an actor in contentious politics of provisions is highlighted by how the Mozambican government turned to the radio to gather information on public sentiment, as popular mobilization for another *greve* (strike) began to emerge in response to a fresh set of price hikes. This time, the price rises were announced well in advance and in combination with specific mitigation measures, and government officials used radio phone-ins to test popular response to these measures before their official introduction.

In the context of India, too, the media were important actors and targets. The Right to Food Campaign had a sophisticated and comprehensive media strategy. Senior campaign members were regularly featured in the national media to discuss opposition to features of the draft National Food Security Act. The Madhya Pradesh case study on the Right to Food struggles of the *Adivasi Adhikar Manch* (translated into English as AAM) further brings out the intricacies of this media strategy, revealing in finer detail how allying with the media became critical to the success of the struggle. Although the media did not always cover AAM's mobilization and public action, they coordinated to raise the issue of malnutrition and the failure of government response, partly through close association with the media activist organization Vikas Samvad.

Conclusions: why media framing of food riots matters

This chapter took a media frames approach to the analysis of how food riots and other subsistence protests were covered in the international and national print media during the global food crisis (during the period 2007–2012). A conclusion by the international media framing of the food riots pointed to a global level in the politics of provisions, with debates and negotiations over the role for multilateral action to address global food security, emerging in direct response to the globalized spread of the protests. The language of 'the threat' posed by food riots to global socio-political and economic order is a dominant thread in the international press. The international press also features a great deal of advocacy and opinion on these issues, often bringing the heads of multilateral agencies, the UN, the World Bank, and associated specialist agencies into the global public space to offer their analysis. But protestors' voices are comparatively muted. Their motivations are often flattened out to the universal stimuli of 'hunger' and 'anger'. But there are also finer-grained analyses in the international press. Many of these protestors' voices spoke in broadly moral economic language, of rights to food and the responsibilities of public actors.

'Food riots', it turns out from analysis of both the international and the national media coverage, can as often be an accusation of state failure as it is a description or analysis of political events. A national press rarely uses the term to describe events at home, but often uses it to describe similar events abroad. It is not clear why this should be the case. However, it may reflect the way in which knowledge of the wider struggles within which such events are nested complicates how people imagine such events. If you are writing for a Dhaka audience that already knows how little garments workers are paid, how hard they work, how lavishly factory owners live, how expensive food has become, and that the garments workers frequently protest about these issues in combination, it makes no sense to label it a food riot, with its suggestions of the moment of a hungry, angry mob.[8]

Protestors' voices were louder in some of the national media, with their more contextualized framing, than in the international coverage. It no doubt matters that in particular for the vernacular press, the audience for such news would include groups similarly affected by the food price crisis, and who shared other characteristics with protestors, including geographical proximity. Even then, descriptions of violence often neglected state violence while emphasizing that of protestors.

A methodological conclusion is that while highly imperfect registers of the range of contentious politics around food crises, media coverage of 'food riots' offers valuable insights into the kinds of subsistence protests that enter the space of public opinion via media coverage. That there are biases is obvious and inevitable. But international media coverage makes it possible for global public opinion to witness the global spread of conflict arising from food insecurity. This 'scale-shift' (McAdam, Tarrow, and Tilly 2008) from localized grievances to similar struggles across the globe was dramatized in the international press in apocalyptic terms. But while dramatic, these documented shifts are unlikely to be an accurate count of all such protests. Biases against rural and small-town areas mean national media almost certainly underestimate their total frequency, particularly where they are so frequent as to lack any newsworthiness at all. Newsworthy protests may be those that directly affect the interests of the elite or the audience, a subset of the total; these are in turn likely to more important political events because they draw powerful groups into their struggles. If there are reasons to believe media coverage yields only a subset of the most important 'food riots', it was enough during the global food crisis of 2007–2012 to document a spread of important grievances about food prices across the globe. The very globalness of this wave of protests in turn created public space for deliberation over key issues in global food governance: climate change and renewable energy, trade and financialization, and industrial vs. smallholder agriculture.

It is difficult, however, to study the mechanisms of contentious food politics using media sources. Our adaptation of the Tilly (2008) approach was to try to situate individual events within their wider set of interactions between protestors and public authorities. The content of the national media was rarely detailed or longitudinal enough to enable significant analysis of these interactions. To complement the newspaper coverage analysis, we also undertook qualitative case study approaches using primary research to deepen our understanding of the means and motivations and repertoires of subsistence protestors. It was through the interrogation and triangulation enabled by the use of these mixed methods that we found it possible to uncover some of the mechanisms shaping subsistence protests and other forms of contentious food politics during the global food crisis.

Notes

1 See also Hossain and Kalita (2014) and (N. Hossain 2009) for instances of moral economy thinking during the recent global food crisis.
2 It should be noted that some articles appeared to use the term euphemistically, perhaps to attract attention in a period during which food riots had become common. For instance, the official response to teenage vandalism of a burger chain in Edinburgh, not long after the wave of global protests about food prices, was reported by the Daily Record as 'Yobs Not Locked Up Over Food Riot' (September 19, 2008, p. 21).
3 'Yobs Not Locked Up Over Food Riot', *Daily Record*, September 19, 2008.
4 www.thedailystar.net/news-detail-118957 [accessed February 15th 2017].
5 www.thedailystar.net/news-detail-32038 [accessed February 15th 2017].
6 http://in.reuters.com/article/idINIndia-29970920071012 [accessed February 15th 2017].
7 http://timesofindia.indiatimes.com/india/Cong-hits-back-assails-Left-for-Bengal-food-riots/articleshow/2481814.cms [accessed July 13 2017]; timesofindia.indiatimes.com/india/Panel-blames-corrupt-PDS-for-Bengal-food-riots/articleshow/2639176.cms [accessed February 15th 2017].
8 For critical analysis of media coverage of recent food riots, see (Sneyd, Legwegoh, and Fraser 2013), whose analysis has been very helpful in thinking about the media and the politics of food crises. See also (Engels 2014).

References

Abdinur, Mustafa Haji. 2008. "Food Riots Turn Deadly in Somali Capital." *Agence France Presse*, May 5.

Barlow, Karen. 2008. "Food Riots 'an Apocalyptic Warning.'" *ABC Premium News*, April 14.

Benford, Robert D., and David A. Snow. 2000. "Framing Processes and Social Movements: An Overview and Assessment." *Annual Review of Sociology* 26 (1):611–639.

Bohstedt, J. 2010. *The Politics of Provisions: Food Riots, Moral Economy, and Market Transition in England, C. 1550–1850*. Farnham, Surrey and Burlington VT: Ashgate.

Engels, Bettina. 2014. "Contentious Politics of Scale: The Global Food Price Crisis and Local Protest in Burkina Faso." *Social Movement Studies*, May, 1–15. doi:10.1080/14742837.2014.921148.

Entman, Robert M. 1993. "Framing: Toward Clarification of a Fractured Paradigm." *Journal of Communication* 43 (4):51–58.

Holt-Giménez, Eric, and Raj Patel. 2012. *Food Rebellions: Crisis and the Hunger for Justice*. Cape Town, South Africa; Oakland, CA and Boston, MA: Food First Books.

Hossain, Naomi. 2009. "Reading Political Responses to Food, Fuel and Financial Crises: The Return of the Moral Economy?" *Development* 52 (3):329–333. http://dx.doi.org.ezproxy.sussex.ac.uk/10.1057/dev.2009.34

Hossain, Naomi, and Devangana Kalita. 2014. "Moral Economy in a Global Era: The Politics of Provisions During Contemporary Food Price Spikes." *Journal of Peasant Studies* 41 (5): 815–831. doi:10.1080/03066150.2014.895328.

Hossain, Naomi, Luis de Brito, Ferdous Jahan, A. Joshi, Celestine Musembi, Biraj Patnaik, Michael Sambo, Alex Shankland, Patta Scott-Villiers, Dipa Sinha, Devangana Kalita and Nicholas Benequista. 2014. "Them Belly Full (But We Hungry): Food Rights Struggles in Bangladesh, India, Kenya." Synthesis report from the DFID-ESRC Food Riots and Food Rights project. Brighton: Institute of Development Studies. www.ids.ac.uk/publication/them-belly-full-but-we-hungry-food-rights-struggles-in-bangladesh-india-kenya

IRIN. 2008. "Burkina Faso; Food Riots Shut Down Main Towns." *Africa News*, February 22.

Katz, Jonathon M. 2008. "Haitian President Calls for Tax Cuts on Imported Food to End Food Riots." *The Associated Press*, April 9.

Khanna, Akshay. 2012. "Seeing Citizen Action Through an 'Unruly'Lens." *Development* 55 (2):162–172.

Koop, David. 2008. "Behind the Food Riots: A Debate on How Best to Farm." *The Associated Press*, May 11, sec. Business.

Matthes, Jörg. 2009. "What's in a Frame? A Content Analysis of Media Framing Studies in the World's Leading Communication Journals, 1990–2005." *Journalism & Mass Communication Quarterly* 86 (2):349–367.

McAdam, Doug, Sidney Tarrow, and Charles Tilly. 2001. *Dynamics of Contention*. Cambridge: Cambridge University Press.

———. 2008. "Methods for Measuring Mechanisms of Contention." *Qualitative Sociology* 31 (4):307–331. doi:10.1007/s11133-008-9100-6.

O'Grady, Sean. 2010. "Russian Wheat Export Ban Threatens Higher Inflation and Food Riots." *The Independent*, August 6, First edition, sec. Business.

Patel, Raj. 2012. *Stuffed and Starved: The Hidden Battle for the World Food System*. 2 Rev Exp ed. Brooklyn, NY: Melville House.

Reuters. 2010. "Economic, Political Risks Float in Pakistan Floods: Possible Scenarios Could Include Food Riots, Government Takeover or Fiscal Collapse." *The Gazette*, August 19.

Scheufele, Da. 1999. "Framing as a Theory of Media Effects." *Journal of Communication* 49 (1):103–122. doi:10.1111/j.1460–2466.1999.tb02784.x.

Sneyd, Lauren Q., Alexander Legwegoh, and Evan D. G. Fraser. 2013. "Food Riots: Media Perspectives on the Causes of Food Protest in Africa." *Food Security* 5 (4):485–497. doi:10.1007/s12571–12013–10272-x.

Thompson, E. P. 1971. "The Moral Economy of the English Crowd in the Eighteenth Century." *Past & Present* 50:76–136.

———. 1991. *Customs in Common*. London: Penguin.Tilly, Charles. 1974. "Food Supply and Public Order in Modern Europe." In *Food Supply and Public Order in Modern Europe*, edited by Charles Tilly. Princeton, NJ: Princeton University Press.

———. 2008. "Describing, Measuring, and Explaining Struggle." *Qualitative Sociology* 31 (1):1–13.

United Press International. 2011. "Morales Aborts Visit Amid Food Riot Fears," February 11. upi.com.

4 Food riots in Bangladesh?

Garments worker protests and globalised subsistence crises

Ferdous Jahan and Naomi Hossain

Introduction

This chapter explores the politics of provisions in Bangladesh through an analysis of the so-called 'food riots' of 2008. On the face of it, poor, malnourished, mal-governed Bangladesh was a clear candidate for food riots during the global food crisis, which the international media duly reported in April 2008. And there were indeed subsistence-related struggles, which resonated loudly with the prevailing moral economy. But in intent, form, and repertoire, these struggles rarely targeted failures of food marketing systems or demanded action by public authorities, and in no respects resembled what social history has taught us to think of as 'food riots'. Although triggered in part by food price rises, these protests reflected a strong sense of subsistence crisis among the protesting garments workers, and were part of a longer series of struggles over wages by this emergent, globalised precariat.

What do these protests tell us about the politics of provisions, or the struggle between elites and masses to secure basic protection against crises of subsistence? This chapter argues that Bangladesh has, for reasons of political economy, developed a comparatively responsive state system of food security since the devastating famine of 1974. Its food security system marries moral with political economy, guaranteeing that the interests of the hungry masses and those of the ruling elite align, at least to the extent of providing protection against crises of subsistence and survival. For this reason, Bangladesh is unusually good – for a country at its level of economic and institutional development – at providing such protection (Hossain 2017). But its precociousness with respect to national food security is not matched by forward thinking on industrial relations: wages have been kept so low that the steep rise in global food prices from 2007 caused a subsistence crisis among workers in the flagship garments industry, adding impetus and powerful moral economic resonance to an ongoing wage struggle. The struggles cast by the international media as 'food riots' are more correctly viewed as the wage struggles of an emerging class of globalised, precarious industrial labour. These wage struggles were not the contentious politics of a population whose state had yet to develop the food security apparatus necessary to safeguard against major crises of subsistence, but reflected instead the exclusion of that emergent precariat from the protection available to others. The prominent struggles of Bangladesh's garments workers during the global food crisis point to what John Bohstedt terms the 'politics of provisions' as they play out in the recent, globalised subsistence crises of developing countries (Bohstedt 2016); this is the nature of contemporary grievances around food, and the opportunities affected populations have to mobilise 'collective bargaining by riot'.

The chapter is organised as follows. It first describes the food crisis of 2008 as it unfolded in Bangladesh, and examines some of the official responses, both short- and longer-term. Then it explores the politics of provisions during this period, looking at how the moral and political economy of food price crises played out in protests and other forms of popular and elite advocacy for action to correct failing food markets. It takes a closer look at how garments workers' struggles were reported on and perceived, and at garments' workers own views on the meaning and purpose of their protests. The chapter concludes with a discussion of what the garments workers' protests tell us about the politics of provisions in Bangladesh, both its broad effectiveness in achieving protection without need for widespread protest, at which groups were always excluded from such protection, and had the opportunities and means to mobilise to protest.

The global food crisis in Bangladesh

The onset of the global food crisis in 2008 was, for Bangladesh, a matter of triply unfortunate timing: excess flooding and Cyclone Sidr in 2007 had reduced the harvest by almost two million metric tons (World Bank 2010); national foodgrain reserves had been run down, in line with policy recommendations by the World Bank (Islam 2012; Chowdhury, Farid, and Roy 2006); and the country was being governed by an unelected caretaker government (Ahmed 2010). As the global financial crisis unfurled around the world, workers in the flagship readymade garments industry had already been protesting low wages and breaches of labour rights since 2006 (Rahman and Langford 2012; Siddiqi 2017). The high price of food could only exacerbate their grievances (Rashid, Hasan, and Hossain 2012).

The effects of the food crisis

The high price of food caused a real, if short-lived, subsistence crisis for the majority of Bangladeshis. When global food prices soared in 2008, the country was classed among the largest developing countries at risk of a major food crisis (De Janvry and Sadoulet 2008). As late as 2005, 40 per cent of the population lived below the country's poverty line, a quarter in extreme poverty (World Bank 2008). Almost half the children were underweight and 12 per cent severely underweight in 2007 (NIPORT 2009). And despite progress towards rice self-sufficiency (Murshid *et al.* 2009), Bangladesh was 'substantially deficit in food grains' when world prices spiked in 2007, partly due to Cyclone Sidr (M. Hossain 2010: 175). In usual years, the domestic grain production shortfall was met by Indian imports, averaging around 5 per cent of total consumption annually

(World Bank 2010). However, the effective ban on Indian rice exports at the height of the global crisis saw consumer rice prices double in 15 months, from Bangladesh Taka (BDT) 18 per kg in January 2007 to BDT 35 per kg in April 2008 (M. Hossain 2010); see (Figure 4.1).

The effects of the rising price of rice were wide and immediate. Nationally, half of all spending was on food *before* prices rose. People started eating less and less diverse diets, with lasting adverse effect for infants and small children (Save the Children 2009; Thorne-Lyman *et al.* 2010; Sulaiman, Parveen, and Das 2009). Children were withdrawn from school (Raihan 2009), and people endured great physical hardships and indignities in order to meet basic needs (Rashid, Hasan, and Hossain 2012: 75).

It was not only the poorest who suffered, but people with regular but low-paying jobs, factory workers, and the lowest grades of the civil service or 'fixed income' groups. Unlike domestic staff, they could expect no patronage from the better-off; and unlike self-employed transport workers or vendors, they found it difficult increase their working hours or bid up their earnings. Even middle class folk were seen queuing to buy the subsidised (often poor quality) rice offered for sale in the Government's Open Market Sales (OMS) scheme (Khan and Wadud 2010). Toiling away in the factories for 10 or more hours a day, six days a week, garments workers found it particularly difficult to take advantage of the OMS.

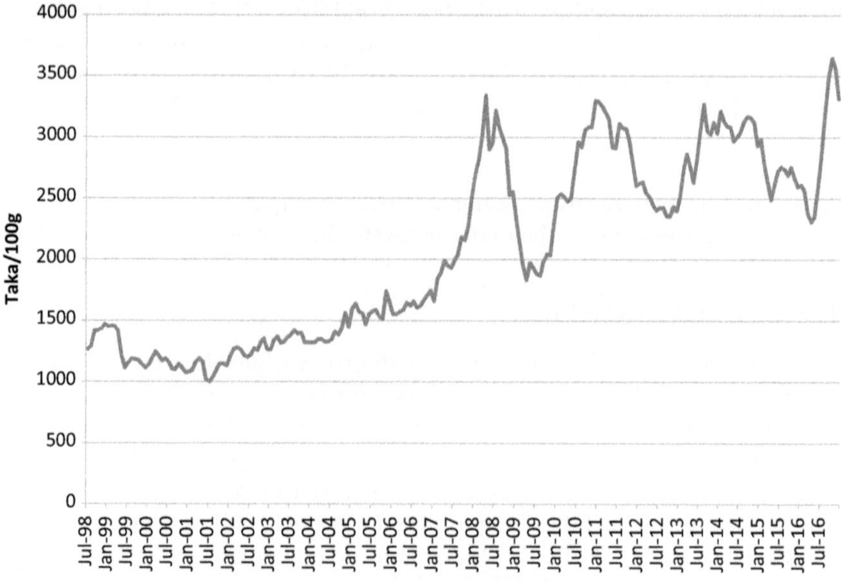

Figure 4.1 Wholesale price of coarse rice, Dhaka 1998–2016

Source: FAO GIEWS www.fao.org/pricetool – accessed 27 February 2017

Bangladesh, Dhaka, Rice (coarse- BR-8/ 11/ Guti/ Sharna), Wholesale

Garments industry wages have been low, and until recent minimum wage rises, had stagnated in real terms since the beginning of the century. The price spike of 2007–2008 left garments workers' at 'starvation levels of living' (Muhammad 2011: 25), eating poorly, and sharing overcrowded, unsanitary housing, even while working long and arduous shifts. In real terms, although garments workers saw rises in the minimum wage rate in 2006 (from Tk 930 to Tk 1,662 per month, the first rise since the wage was established in 1994) and again in 2010 (to Tk 3,000), food price inflation meant their real pay stayed more or less stagnant throughout the period (Ahmed and Nathan 2014; Rahman 2010; Muhammad 2011). In other words, workers struggled at great risk to life and limb, only to remain on the lowest wages in Asia (Moazzem and Raz 2014).

The international media broadcast news of food riots in Bangladesh in April 2008, amidst a wave of similar protests in similarly poor and hunger-prone countries worldwide. In Bangladesh, these were framed as food riots (The Times of India 2008), food price riots (Al Jazeera 2008), and strikes or riots by garments workers over food prices (Ramesh 2008) in the second week of April 2008, when the price of rice also peaked. Bangladesh appeared in lists of countries experiencing food riots (for instance, CNN 2008; Schneider 2008). Bangladesh made several lists of countries that experienced food riots in 2008,[1] shorthand for the kinds of places likely to suffer when food crises strike.[2]

Reaffirming national food security[3]

The global food crisis of 2008 came as a sharp shock to the ruling class and the state, and immediate and sustained action was taken. A key lesson from the 2008 food crisis appears to have been about food sovereignty: that the government of Bangladesh must protect its power to influence the stability of domestic food prices. Under conditions of global food price volatility, it cannot rely wholly on international trade to achieve that, as trading partners may also be pursuing their own food sovereignty goals at the expense of their neighbours'. This implies the continued need for a strong and capable public food distribution system with the agility to respond to crises. It also implies the need for stronger capacities for food storage, which are being addressed through the Modern Food Storage Project.[4] It also implies the need for a public food distribution system that enables the government to combine the need to rotate its public foodstocks with its goals of poverty and hunger reduction and food and nutrition security. The state has relied increasingly on getting food markets to work better to ensure stable rice prices (see Ahmed, Haggblade, and Chowdhury 2000), and Bangladesh continues to view international trade as crucial to its food security: when India closed its borders in 2008, Bangladesh started to diversify its imports with new trading relationships with Myanmar, Pakistan, Viet Nam and Thailand (World Bank 2013).

But while the overall thrust of food policy management has been to enable markets to work more effectively, the social and political costs of food price spikes mean a continued role for public intervention, particularly food price stabilization. Some of the key achievements with respect to food crisis management have included interventions that have averted market failures due to hoarding or other collusive behaviours by major market actors (see, for instance, Osmani 1991; Banerjee *et al.* 2014). The success of OMS and other fair price programmes depended on the state having foodstocks to release into the market, to stabilise prices, reassure consumers and traders that food prices would not go too high, and ensure food affordability and access. But in 2007–2008, the then-caretaker government failed to import enough grain to be able to make an appreciable difference in the market (Ali *et al.* 2008). In part, this was due to the decline in grain storage capacity: in the 2000s, the government of Bangladesh had reduced reserves with the aim of relying on the global market. By 2007–2008, only half a million metric tonnes were distributed through the Public Food Distribution System (PFDS) (Raihan 2013), when the National Food Policy declared a million tons of public food stocks as its objective. After 2008, distribution through the PFDS rose to 2.29 million tons; at its peak during the 2007–2012 crisis period, public food stocks amounted to 2.77 million tons, of which 1.6 million were to be distributed through the safety net programme (World Bank 2013). A new programme of warehouse construction featuring improved grain storage facilities has been supported since the 2008 crisis by the World Bank, which appears to have acknowledged that under these new conditions of price volatility, it is appropriate for the government to have a 'renewed focus on ensuring a higher level of public stocks . . . a consequence of hikes in international rice and prices and disruptions of rice imports in 2007–2008' (World Bank 2013: 109).

Social protection

The global food crisis also helped persuade the government of the need for stronger social protection provision, noting in particular the exposure of the urban working classes to the global economy, without the customary protections:

> [W]ith informal safety nets eroding, newer risks emerging from rapid processes of urbanization and global economic integration, and, stronger assertion of mitigation demands from a democratizing polity, a holistic re-thinking on the direction, scope and design of safety net policies in particular and social protection policy in general has become necessary.
>
> (GoB 2012: 165)

The 2008 food crisis prompted the establishment of at least one important new social protection scheme, the Employment Guarantee for the Poorest. Although this was originally introduced under the caretaker government in 2008, the incoming Awami League government rebranded and expanded the scheme in its first term. The National Social Security Strategy stated that 'Despite numerous administrative problems, these programmes [FFW, the EGPP, and the OMS] have helped cushion against the impact of food price shocks' (GED 2015: 41). However, it seems that provision falls well short of need: a 2009 nationally representative survey of household shocks and coping mechanisms found that 29 per cent of respondents had access to a public safety net, but that these ranked fourth in the list of sources of help with the crisis (Santos *et al.* 2011). Analysis in the wake of the 2009 Cyclone Aila found that 90 per cent of relief (food, cash, water, medicine, etc.) was distributed via state safety net channels and was well-targeted to those affected, but resource levels were inadequate to make a significant difference, and need was far wider than provision (Akter and Mallick 2013; Akter and Basher 2014). Direct social protection against food crises remains somewhat token – gesturing towards rather than guaranteeing food security.

Subsistence protests and civil society action, 2007–2012: who, when, how, and why

Who protested or demanded action, why, and how during the global food crisis? To what extent did their claims hold government to account? What does the pattern and nature of these contentious events tell us about the politics of provisions in Bangladesh more generally?

To address these questions, we constructed a political events catalogue from a careful digital and/or hand search of three of the most influential daily newspapers in Bangladesh: the English language *Daily Star*, the Bangla *Ittefaq* and the highest circulation Bangla *Prothom Alo*. A search protocol was developed, tested, and refined, and a database developed to enable systematic quantitative and qualitative analysis of all reported events in which food security or related issues of subsistence were in contention during 2007–2012, the period of both major global and domestic food price spikes. To ensure coverage of a sufficiently wide range of contentious events to capture the quality and tenor of the politics of food crisis, claims-making in relation to food consumption (e.g., demands for fair priced food, price controls, subsidies, food relief) were included as well as in relation to food production (fertiliser, costs of irrigation, other agricultural inputs), as well as subsistence protests or claims-making activities that explicitly identified the rising cost of living as a source of subsistence crisis. The experience of constructing a political events catalogue showed us the great variability of coverage of such events, including urban bias. After selecting from several thousand newspaper articles, a total of 168 newspaper articles covering 86 discrete events – some of them connected in protracted struggles – were analysed in detail.

Although worldwide, this was a highly febrile, protest-rich period (Ortiz *et al.* 2013; Barnett 2011), contentious events relating to food or fuel were, perhaps surprisingly, few in number in densely-populated, low-income Bangladesh. This may have been due to the official ban on political demonstrations during the caretaker government, which was in place for the early part of the food crisis, in 2007 and 2008. The 86 contentious events relating to the food crisis reported on in 168 articles, many of them non-violent or 'civil' in nature contrasts with estimates of political violence, chiefly between political party supporters, of around 2,423 episodes of 'political conflicts' between 1991 and 2002 (Moniruzzaman 2009), and 2,859 of political violence, including terrorism, between 2001–2006 (J. Alamgir 2011). But while food crisis-related contention was limited in scale, it clustered around moments when retail prices of basic foodstuffs were peaking. This suggests that those who protested were persuaded of the seriousness and justice of their protests, and/or disinclined to believe the public authorities would suppress them. And that they were few in number does not mean the protests were insignificant; indeed, their rarity may tell us something important about the moral and political economy of food crises in that context.

Contentious actors and actions, 2007–2012

Table 4.1 lists all groups and organisations mentioned in all news articles about all types of contentious events relating to the food crisis in this period. These included riots, protests, and demonstrations, as well as more 'civil' or organised events staged by civil society organizations ('human chains', petitions, roundtable discussions) that also fell outside the formal political process. The main political parties took up issues of the cost of living in the period, particularly while they were in opposition, and both parties when in opposition led demonstrations to raise the issue. A small number of civil society groups also mounted actions, but these were usually more polite or formal organizational means of advocating action than protests. Some involved stylised repertoires of contentious politics such as demonstrations or actions staged for the benefit of the media, such as the token hunger strike. However, organised labour and worker welfare organisations were more prominent than party political actors. Garments workers' protests led the fray: almost 40 per cent of all groups mentioned in relation to subsistence protests were garments workers or their organisations. The next most important category was farmers and fisherfolk and their welfare groups/councils, with almost one-fifth of all mentions. Other groups with a significant presence in the protests were other labour organisations, including those of low paid public sector workers' unions and groups. The rural poor, including destitute women and people affected by disasters or demanding relief, comprised only 6 per cent of all mentions.

Newspaper reports of grievances to do with rapid food price rises were less frequent during the 2010–2011 spike than in 2007–2008. This may reflect that in real terms, the second price spike was less severe than the first, and wages had risen for many (Zhang *et al.* 2013). It is possible that the drama of price spikes had subsided by 2010, and that people had adjusted their lives and expectations to the 'squeeze' of higher rates of inflation (Hossain, King, and Kelbert 2013). In addition, protections against food crises in the form of the national foodgrain reserve and direct food transfers and subsidies were more in evidence and quicker to kick in than they were in 2008.

Whatever the precise mix of factors, those who protested in 2010–2011 were still aggrieved about high food prices, but the objectives and triggers for subsistence protests were now even more closely focused on wages and labour rights than had been the case in 2008. It may not, then, be a coincidence that there were 30 episodes of garments industry violence in the six months between August 2010 and January 2011 (Rahaman *et al.* 2011), at the height of the

Table 4.1 Protesters and petitioners as described in articles on contentious events

Political parties and actors		*Social movement/civil society groups*		*Occupational groups*		*Trades unions/labour organisations*		*Other*	
Main parties	10	Consumer Association of Bangladesh	2	Garments workers	35	Worker welfare associations and councils	13	Villagers/ poor people/destitute women	10
Leftist parties	7	Student and cultural groups	4	Farmers/ fishers	17	Blue-collar labour unions/ federations	9		
Local political representatives	4	Think tanks	2	Economists	2	Bangladesh land port workers' league	1		
Parliamentary Committee on Food and Disaster Ministry	1	Development NGOs	2	Intellectuals, professionals, journalists	2	Garments workers	27		
		Regional or area organizations	2	Fertiliser dealers	1	Public sector worker unions/organisations	7		
		employers and business associations	3	Agricultural land owners	1	Coal miners	1		
		Local citizen groups	2						
	21		17		58		57		10

Note: this table takes the news article rather than the event as the unit of analysis. Figures used here are of citations for each time the group was mentioned and so provide a sense of how the events were covered in the media, not of the extent of activism by each group.

second food price spike; but by this time, perhaps as a result of their success at having bid up minimum wages in 2010, their attention was on the minimum wage, rather than on food prices (which appeared to continue to rise) or on the marketing of rice, which proceeded as it always had. Other groups similarly focused on their rights to earn a living wage rather than on what that wage could afford them, despite remaining aggrieved by high food costs. In early 2011, as food prices spiked the second time, Dhaka rickshaw-pullers told researchers they felt they needed to protest to alert the attention of the government to unlivable inflation. Yet when members of their profession rioted three weeks later, they cited the closure of their usual routes, rather than food prices, as their grievance (Rashid, Hasan, and Hossain 2012). This suggested that inflation had become normalised or de-politicised, and that, although still a major public grievance, high prices no longer afforded the political opportunity for protest they once did.

Several of the more 'civil' events to advocate for public action on food prices addressed similar points to subsistence protestors themselves. At a pre-budget consultation meeting, a group of respected economists denounced governmental inaction to address food prices, noting that 'economic indicators could not feed people' (*Daily Star* 2008). Such awareness of and/or proximity to the plight of those living with poverty and food insecurity is, for political/historical reasons, characteristic of the elite-mass relationship (Hossain 2005). It suggests a degree of cross-class consensus with respect to the moral economy, or views about the responsibilities of public authorities to protect the population against the failures of food markets.

National government, in particular the executive and line ministries (e.g. food), were a key object of most major protests and advocacy events, but sub-district and lower-level officials and elected representatives were targets of more localised protests. Global policies and actors were rarely objects of protest, other than a handful of events protesting actions of corporations and large companies, including foreign investors (e.g., the China Contractual Authority). Fully one-third of all targets of protests mentioned in all the articles were garments factory owners and their association, the Bangladesh Garments Manufacturers and Exporters Association (BGMEA). Through these indirect routes, the subsistence protests targeted transnational actors and institutions.

Figure 4.2 provides a basic count of the types of action described in the article (in some cases more than one action was mentioned). All are from the *Dainik Ittefaq*, which provided a relatively rich archive in the vernacular. Reflecting the dominance of industrial conflict in the sample, the description of the events and the actions taken are substantially more often unruly and disruptive (demonstrations, agitations, sieges, marches/processions) than civil (hearings, human chains, memoranda, discussions, petitions). This pattern extends also to the violent nature of these events: in some 30 events, or more than one-third of the total, vandalism of factories or vehicles and threats of violence occurred. The impression given by the news coverage is also that these were usually unorganised events – riots and agitations rather than organised lobbying or civil action: out

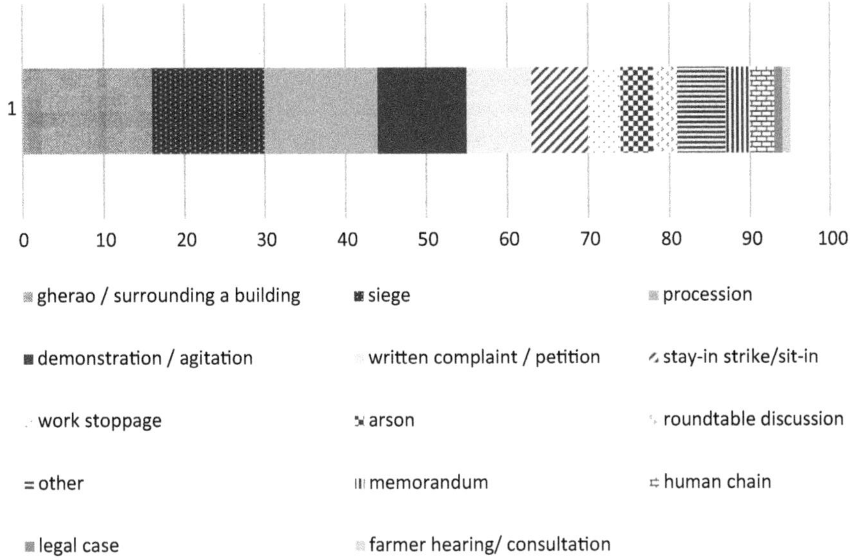

Figure 4.2 Repertoires of protest

Note: n=95 because more than one action could be recorded during a single event

Source: Political events catalogue, *Dainik Ittefaq* archive

of 84 events, 47 were described in terms that suggested spontaneity and lack of organisation, and 36 were characterised as organised or planned events.

We turn now to a closer look at how garments workers themselves perceived their grievances and opportunities for mobilization during this period of subsistence crisis.

Wage struggles and subsistence protests among garments workers, 2002–2012

This chapter draws on the case study of garments workers' protests in the peri-urban industrial area of Savar, north of Dhaka, the April 2008 episodes of which had been reported by the international press as 'food riots'. Because of the sensitivities of researching such events, the case study research involved preliminary assessment and observational work and then a period of building trust and rapport with workers; only then were focus groups and interviews conducted. Focus group discussions used prepared protocols with protest participants, workers from the area who had not themselves participated in the protests, and local market stallholders and shopkeepers, while interviews with a small number of journalists, leaders and activists and officials were also undertaken to fill in gaps and triangulate findings.

Savar has been the focal point of industrial unrest since the 2000s, owing in part to the concentration of garments factories and the presence of the Dhaka Export Processing Zone (EPZ) in this sub-district, 25 km north of Dhaka. Savar has more recently become associated with exploitation in global export industry, as the infamous site of the Rana Plaza building collapse which killed 1,130 workers and injured almost 5,000 more in April 2013 (for accounts of which, see Karim 2014; Motlagh and Saha 2014). More than any protest, the tragedy and horror of that event exposed the chronic precariousness of workers' lives, and their dependence on dangerously exploitative jobs to earn the barest subsistence.

Moving targets and shifting repertoires in the garments workers' struggles

Garments workers offered a timeline of the struggles in their industry that placed the so-called 'food riots' of 2008 in perspective as a comparatively minor, if distinctive, moment in a longer struggle that had evolved over time, in relation to the political opportunities and grievances workers faced. Workers in Savar linked the 2008 and 2010–2011 struggles (featured in the political events catalogue, and referred to previously) to a longer sequence of struggles for adequate pay and conditions that dated back to 2002, and continued throughout the period covered by the research. Protests were initially motivated by demands for increased wages and against the illegal practice of terminating workers' employment without justification. In 2004, workers told us that the main protests spread out across factories in the area after an illegal sacking which had in turn been triggered by demands for trade union rights in a sweater factory. The police and local gangsters were called in to quell the protest, so organisers soon learned to organise outside the factory walls. These struggles were often fruitful: in one factory, workers described how they had been permitted to organise after a demonstration entered the factory.

By 2006, protests had shifted from targeting the management of individual factories to making demands of industry leadership and the state, through a movement focussed on the single and clear demand of raising the minimum wage. Since its establishment in 1994, the basic monthly minimum wage had been stagnant at BDT 930 (in January 2017 exchange rates, USD 11.74) (Ahmed and Nathan 2014).[5] In 2006, the struggle that emerged was large in scale and spread rapidly through the Export Processing Zones (EPZ),[6] despite repressive tactics by the police. The 2006 struggles were widely seen as a success for the workers, as the basic monthly minimum wage rate was raised, for the first time since its establishment, to BDT 1,662 (currently, USD 20.98). As the workers explained it, these protests were generally self-organising, without involvement by external actors. They adopted non-violent means to protest, viewing acts of violence as committed by 'outsiders' – professional agitators, not workers. The major political parties offered workers no support, although leftist political organisations backed them. Workers began to learn from their successes.[7]

There was a qualitative change in the political space for workers' protests after 2007, as external actors – workers' rights organisations, leftist coalitions and groups, and activist and civil society organisations, several with roots in the garments workers' struggles – began to engage with their struggles. Through this engagement, workers' repertoires widened to include dialogue with factory owners and industry leaders, information campaigns through leafleting about workers' demands, and the use of peaceful human chains. Even with the involvement of new actors, the main source of mobilisation was stimulated by the 'in-charge' workers who commanded workers' respect. Experienced leadership of 'in-charge' workers was used to organise a large group of workers, who would then spread the word by mobile phones and other means. In some factories, even this relatively flat leadership was absent, and groups decided on actions without clear leadership. The involvement of external actors in supporting the workers' protests did not reflect in the mobilisation. The site of mobilisation remained the factory.

Growing inadequacy of wages during the period of rapid food price rises from 2007 led to relatively spontaneous and unorganised responses by workers in Savar. One reason was that, by 2007, workers were accustomed to taking to the street, and to taking factory-based action. Actions that started in one factory quickly spread to others, as news of wage increases were shared. The 'movement' was sporadic and stop-start, and never sought a long-term solution to their problems: direct action stopped whenever a raise was realised, and re-started when increased wages were found to be inadequate.

Throughout, the workers' struggles had no major-party political support (no doubt reflecting the importance of garments factory ownership among the political elite). Small leftist political parties joined in actions later during this period, but played no significant role in organising workers. A mutual lack of interest between the workers' struggle and party politics reflected the workers' distrust of political parties of all stripes. One male worker in a focus group with (non-active) male garments workers noted that 'we do not believe in banners (of political parties) – that will compromise our movement' (focus group discussion on July 23, 2013, Savar, Dhaka).

The moral economy of garments workers' protests

Over time, garments workers framed protests in different ways, in part depending on the wider political context, and in part on the political education gained from (at times) successful organizing. However, they continued to lack important party political backing and articulated their grievances and demands without reference to leftist or popular groups. Despite the presence of some established leftist groups (such as the workers' wing of the Jatiya Samajtantrik Dal and the Communist Party of Bangladesh) at some protests, workers' struggles did not noticeably make common cause with other urban groups on fixed low incomes (such as other industrial or low-paid public or private sector workers). This may partly reflect the central significance of garments labour to the economy as a whole: this labour-intensive industry alone contributes 80 per cent of export earnings

(Bangladesh Bank 2015; MoF 2015), and employs more than half of all industrial labour (BBS 2013). It may also reflect the historic co-optation of organised labour by party politics, and the tendency for Bangladeshi garments workers to attract support from international labour interests and groups (see Siddiqi 2017).

However, compared to their early days, when protests almost universally targeted factory owners for wage rises, workers began to frame their struggles in terms of subsistence rights, flagging issues such as the rising cost of living and the minimum wage, and basic labour standards. This was particularly noticeable during the food crisis period, and for the first time, garments workers' concerns appeared to be resonating with those of not only the low-income rural and urban masses, but also the 'respectable' middle and lower-middle classes, all of whom were struggling with the sudden price rises to greater and lesser extent.

Some workers used the term a 'fair wage' to talk about their demands for higher pay. Workers clarified that such a wage afforded not only a basic living standard – not merely enough rice to stave off hunger and maintain factory functionality – but enough scope to invest in the future, for a decent standard of social reproduction. Speaking to us in 2013, when food prices were still believed to be high and the period of the price spikes were vividly remembered, people noted the strong sense of unfairness that their hard labour failed to afford them even basic needs. 'M.H.', a married 40-year-old textile dyeing factory worker based in the EPZ in Savar, explained that, 'Now the ongoing movement (*andolon*) is for fair wages. If government were to give those fair wages, we would see that everywhere people could meet their needs.' Workers spoke about how rising living costs made it difficult for them to remit money to their families in the village – to do so is not only a source of pride but also an important source of support for ageing parents or families left behind.

Price controls were widely agreed upon as one of the key responsibilities of the government. The established idea was that the government can and should set a fixed price or control prices that people can afford to pay. The idea of a fair price or a fair wage refers to a social order in which people can manage their needs to a reasonable standard of living. In the protests in 2008 in particular, garments workers used the slogan '*dam komao, banchte dao*' ('reduce prices, let us live'). Labour leaders noted that food prices became a central part of their agenda at this time, because the rises made it even harder for workers to maintain even basic dietary standards, let alone other expenditures. By 2011, however, even though the political events catalogue identified other, smaller populations of (other, non-garment) factory and other workers protesting high food prices, garments workers' protests were framed more in terms of wage rises. This reflected their successes at getting raised minimum wage levels in 2006, 2010, and (after the fieldwork was completed) 2013.

Food security for garments workers

To label workers' struggles 'food riots' is to misrepresent their meaning; however, the food crisis created distinct political opportunities for the workers to dramatise their grievances around fair pay and conditions. At the same time, exploitatively low wages – workers' 'forced underconsumption' (Araghi 2009) – were part of the same global economic system that led to the food price crisis. In brief, workers' wages had always been unliveably low, and 'the end of cheap food' (Moore 2014) meant that in real terms, they declined further, and sharply so.

That policy elites recognised the moral and the political-economic significance of the garments workers' protests during the food crisis can be gauged from several different responses. The development of greatest significance was the rise in the minimum wage (see Table 4.2). At the time of writing (January 2017), workers' struggles for a higher wage continued. This no doubt reflects both the experience of success in getting higher wages, and the continued inadequacy of the minimum wage in the context of rapid inflation; one 2014 estimate was that the minimum wage should have been set at BDT 8,200, but that the cost of meeting a 'model diet' would have set the minimum wage at BDT 14,900 (Moazzem and Raz 2014).

A practically minor but symbolically significant success saw workers win the right to Open Market Sales (OMS) or state-subsidised foodgrain sales in factories. Specific factory-based sales were necessary because OMS usually take place during working hours, and typically involves queuing for long periods of time. The state had responded by allocating subsidised grains for sale to up to three million workers by May 2009, soon after the newly-elected Awami League government came to power. This was a significant commitment in principle, but in practice was soon abandoned as market prices dropped in the first half of 2009, making the (often poor quality) subsidised OMS grains less attractive (bdnews24.com 2009).

While it never actually improved workers' food security, the deployment of OMS in factories signalled a shift in how garments workers were viewed, suggesting a new recognition of workers' rights to food security as equivalent to those of other urban citizens, the need for special steps to ensure their access was the same as any other citizen, taking into account their particular modes of work and working hours. It also signalled very directly that the state accepted the responsibility or mandate for enabling their access to affordable food. It is possible to argue that this symbolism has gendered significance, in that an estimated 80 per cent of the country's four million garments workers are women. There are no credible estimates of the level of womens' participation in workers' struggles in the 2000s, but although men are assumed to be more active, it is also clear that women

Table 4.2 Minimum wages in the garments sector

Year of revision	*Minimum wage in BDT*	*% increase in real terms*
2006	1,663	8.8
2010	3,000	38.3
2013	5,300	65

Sources: 2006 and 2010 are from Ahmed and Nathan (2014); 2013 estimate of per cent increase is from (Woodruff 2014)

workers have taken a frontline position (Siddiqi 2017; Z. Rahman and Langford 2012). It was partly because of the presumed docility and dependence of Bangladeshi women that wages had been kept low; implicitly, women had husbands or fathers responsible for providing for them, and anyway, they were unlikely to rebel. Among other things, the establishment of OMS in factories signals explicit recognition of women as citizen-workers, or as 'householders' responsible for food security in their own right, and therefore with the right to protection against food crises tailored to their needs.

It should be kept in mind, however, that the Industrial Police were also established in 2010, and that repression has been a significant, if not the most important, response to the workers' struggles (Siddiqi 2017; Z. Rahman and Langford 2012). In addition, neither the Industrial Policies of 2010 or 2016 directly addressed garments workers' wages or food security issues; the 2013 amendments to the Labour Law and other policy initiatives related directly to the after-effects of the Rana Plaza disaster, including the U.S. withdrawal of trade privileges (Kibria *et al.* 2014).

Conclusions

What does this survey of contentious events around the 2008 food crisis tell us about the politics of provisions in contemporary Bangladesh? The chapter noted that the global food crisis hit Bangladesh hard in 2008, its impact aggravated by pre-existing poverty and hunger, two major cyclones, and exposure to global food market volatilities. A series of contentious events, including high-profile violent protests by workers in the flagship readymade garment industry, dramatised and politicised inflationary conditions. But the Awami League government elected in late 2008 took credit for stabilizing the food situation in this turbulent period. Tackling food prices took top billing on the election manifesto that won the party its largest mandate since the 1970s, signalling its intention to handle the crisis at all costs (BRAC University and Institute of Governance Studies 2010). The government shopped further afield for grain imports, expanded open market sales of cheap food, widened the social safety net, and built warehouses for expanded food reserves. When world food prices spiked again in 2010–2011, Bangladeshis were better insulated than many of their neighbours – and better than their grandparents would have been.

Arriving slap-bang in the middle of an unelected, military-backed regime, the 2008 food crisis showcased the wide democratic streak that runs through the politics of provisions in contemporary Bangladesh. A popular political theory heard around the 2008 crisis was that the government was responsible for protecting its people against food shocks; it would do so not out of great love for its people, but because of simple democratic pressures – exerted, if necessary, through direct action, electoral or protest (Hossain and Kalita 2014; Hossain and Green 2011). The coincidence of the tenure of the caretaker government (January 2007 – December 2008) with the rising rice price lost its popularity, and confirmed the idea that military governments were inherently less capable than 'political' governments of protecting the population against food crises (see [Islam 2012]). The U.S. Embassy reported on the worries of an unelected government, highlighting in particular the grievances of garments workers:

> Bangladesh's food crisis threatens the country's fragile Caretaker Government . . . [Eighty] to 90 percent of Bangladeshis surveyed said they were worse off economically under this government than under the previous government. Thirty to forty percent of those surveyed said they had low or extremely low confidence in the CTG, up from 20 to 30 percent in February. Since January, thousands of garment workers have staged protests over high prices and low wages. The government is especially sensitive to discontent in the ready-made garment (RMG) sector, which supplies the country's main export earnings and employs close to 2 million workers. The average wage for a garment worker ranges between USD 45 and 90 per month. Food security has political ramifications in Bangladesh and threatens the stability of an already-weak Caretaker Government committed to hold elections and restore democracy in the Muslim-majority nation.
>
> (American Embassy Dhaka 2008)

As the caretaker government had (somewhat ineffectively) banned political protests, the sound of popular discontent was muffled. Some of the country's superstar economists, a group frequently found on television talk shows and opinion columns, hosted a meeting to discuss Amartya Sen's theory about the relationship between democracy and famine, an unsubtle dig at the unelected regime. Among its many efforts to address the food crisis, the top-ranking general made the politically naïve proposal that Bangladeshis 'should eat potatoes' rather than relying on expensive rice (Unb 2008). So when food prices dropped sharply just after the Awami League government took power in early 2009, this gave credence to the idea that political governments respond faster than military (even though the success reflected the global commodity slump and a bumper harvest more than action by the new government). In general, the failures of the caretaker government were overstated in popular opinion, and the successes of the Awami League government in 2008 slightly lucky. But the popular theory was reaffirmed.

What reasons do Bangladeshis have to believe elected governments will do whatever it takes to protect against food crises? As was noted in the introduction to this chapter, Bangladesh has earned a reputation for having developed relatively responsive systems for protecting its population against the crises of subsistence and survival that come with the country's chronic exposure to natural disasters such as cyclones and food price shocks, such as occurred in 2008 and 2010–2011. The political origins of these systems in Bangladesh themselves originate in the aftermath of an earlier global food price crisis, the OPEC oil crisis of 1973–1974, which pushed up world food prices and subjected starving populations across Africa and Asia to the *realpolitik* of cold war food aid (Gerlach 2015; Rothschild 1976). The 1974 famine in Bangladesh was a humanitarian disaster in which 1.5 million people, two per cent of the population at the time (M. Alamgir 1980), perished. But it was also a disaster for the political elite, contributing directly to the end of the democratic political settlement of the country's independence from Pakistan, the assassination of the nationalist founding father, Sheikh Mujibur Rahman, and his family. This ushered in 15 years of authoritarian rule, and, for better or for worse, paving the way for liberal economic reforms and aid dependence (Hossain 2017). The Bangladeshi elite to this date remain closely attuned to the de-legitimating effects of subsistence crises. When asked why Bangladesh has such an elaborate and relatively responsive system of food security in place, several respondents from the upper rungs of policy and scholarship explained that it was 'because of the famine'. This was sometimes offered as if it were the self-evident underlying architecture of food security and social protection policy, needing no more detailed explanation.

Some protest was likely in the Bangladesh of 2008, as all governments struggled to stabilise prices and protect their vulnerable and the hapless unelected regime fumbled at first with imports and fair price outlets. As the crisis struck, political parties and civil society elites no doubt felt the need to remind the unelected regime of its responsibilities under the moral economy. But the garments workers' protests added new dimensions to these politics of provisions. First, they highlighted the acute vulnerabilities of the country's flagship export industries to not only the low wages that come with a position at the bottom of the global value chain, but also exposure to global commodity price volatilities, and therefore to subsistence crises. Second, they highlighted their exclusion from the protections of the moral economy, which assured that hungry people could at least queue up for cheap or free food. Bangladesh's national food system has, somewhat prematurely, achieved the kinds of security that Tilly noted came with an integrated, openly-traded domestic food market; this reduces the likelihood of riots targeting the marketing of foodgrains (Tilly 1974). But those most exposed to the downsides of global economic integration, the garments workers, also (and partly as a result) have the greatest opportunities to mobilise, and likelihood of success. There were no 'food riots' in Bangladesh in 2008 signalling the pains of adjustment to a market economy, but garments workers' wage struggles were nonetheless a sign of the country's adjustment to its newly industrial position in the global economy.

Notes

1 https://en.wikipedia.org/wiki/2007%E2%80%9308_world_food_price_crisis#Bangladesh (accessed January 10, 2017).

2 For instance, the celebrated essayist Mark Greif opened his 2008 discourse 'On Food' with: 'Food riots have broken out in Haiti, Egypt, Mozambique, and Bangladesh' (Greif, 2016: 37).

3 This section cites grey literature, official documents and newspaper reports. But it relies substantially on interviews with a small number of high-ranking policymakers (government and aid agency) and senior experts and scholars, the aims of which were to reconstruct the policy thinking and practice of the period. Due to the sensitive nature

of some of the material discussed, these interviews have been anonymised to protect the identities of respondents.

4 www.mfsp.gov.bd/project.php; accessed February 16, 2016.

5 Note that workers typically earn considerably more than the basic rate, which is the entry-level pay, and because most also earn overtime. However, take-home pay remains very low in practice, and hourly rates particularly so. See (Ahmed and Nathan 2014; Moazzem and Raz 2014) for discussion of the minimum wages and effective wages in the RMG industry.

6 EPZs are special industrial zones offered for setting up industries by foreign investors. These are protected, closed areas with strict security.

7 See also (Siddiqi 2017; Z. Rahman and Langford 2012).

References

Ahmed, Nazneen, and Dev Nathan. 2014. *Improving Wages and Working Conditions in the Bangladeshi Garment Sector the Role of Horizontal and Vertical Relations*. Manchester: Capturing the Gains, University of Manchester. www.capturingthegains.org/publications/workingpapers/wp_201440.htm

Ahmed, Nizam. 2010. "Party Politics Under a Non-Party Caretaker Government in Bangladesh: The Fakhruddin Interregnum (2007–2009)." *Commonwealth & Comparative Politics* 48 (1):23–47. doi:10.1080/14662040903444491.

Ahmed, Raisuddin, Steven Haggblade, and Tawfiq-e-Elahi Chowdhury. 2000. *Out of the Shadow of Famine: Evolving Food Markets and Food Policy in Bangladesh*. Washington, DC: International Food Policy Research Institute.

Akter, Sonia, and Bishawjit Mallick. 2013. "The Poverty – Vulnerability – Resilience Nexus: Evidence From Bangladesh." *Ecological Economics* 96 (December):114–124. doi:10.1016/j.ecolecon.2013.10.008.

Akter, Sonia, and Syed Abul Basher. 2014. "The Impacts of Food Price and Income Shocks on Household Food Security and Economic Well-Being: Evidence From Rural Bangladesh." *Global Environmental Change*.

Al Jazeera. 2008. "Bangladesh Hit by Food Price Riots." *Al Jazeera*. www.aljazeera.com/news/asia/2008/04/200861422821207204.html.

Alamgir, Jalal. 2011. *Political Violence as a Core Governance Challenge*. Institute of Governance Studies, Brac University, Dhaka, Bangladesh, and Centre for Development Studies.

Alamgir, Mohiuddin. 1980. *Famine in South Asia: Political Economy of Mass Starvation*. Oelgeschlager: Gunn & Hain, Publishers, Inc.

Ali, Amm Shawkat, Ishrat Jahan, Akhter Ahmed, and Shahidur Rashid. 2008. "Public Food Distribution System in Bangladesh: Successful Reforms and Remaining Challenges." From Parastatals to Private Trade: Lessons From Asian Agriculture, the Johns Hopkins University Press, Baltimore, MD and Oxford: Oxford University Press, New Delhi, India, 103–134.

American Embassy Dhaka. 2008. "Response: Impact of Rising Food/Commodity Prices – Bangladesh (WikiLeaks Report No. 08DHAKA485_a)." Unclassified cable to US Secretary of State 08DHAKA485_a. Dhaka: United States Government. https://wikileaks.org/plusd/cables/08DHAKA485_a.html.

Araghi, Farshad. 2009. "Accumulation by Displacement: Global Enclosures, Food Crisis, and the Ecological Contradictions of Capitalism." *Review (Fernand Braudel Center)*, 32 (1): 113–146.

Banerjee, O., T. Darbas, P. R. Brown, and C. H. Roth. 2014. "Historical Divergence in Public Management of Foodgrain Systems in India and Bangladesh: Opportunities to Enhance Food Security." *Global Food Security*, SI: GFS Conference 2013, 3 (3–4): 159–166. doi:10.1016/j.gfs.2014.06.002.

Bangladesh Bank. 2015. "Quarterly Review on RMG: January–March 2015." Dhaka: Bangladesh Bank, Research Department, External Economics Divisions. www.bb.org.bd/pub/monthly/agri_rural_financing/agri-mar2015.pdf.

Barnett, A. 2011. "A Fight for the Future." In *Fight Back! A Reader on the Winter of Protest*, edited by Dan Hancox. London: Open Democracy.

BBS. 2013. "Survey of Manufacturing Industries 2012 (SMI 2012)." Dhaka: Bangladesh Bureau of Statistics, Statistics Division, Ministry of Planning, Government of the People's Republic of Bangladesh. http://bbs.gov.bd/WebTestApplication/userfiles/Image/LatestReports/SMI-%202012.pdf.

bdnews24.com. 2009. "OMS Rice for Garment Workers From Ma . . . bdnews24.com." http://bdnews24.com/bangladesh/2009/03/18/oms-rice-for-garment-workers-from-may-1.

Bohstedt, John. 2016. "Food Riots and the Politics of Provisions From Early Modern Europe and China to the Food Crisis of 2008." *The Journal of Peasant Studies* 43 (5):1035–1067. doi:10.1080/03066150.2016.1170009.

BRAC University and Institute of Governance Studies. 2010. *The State of Governance in Bangladesh 2009: [Entitlement, Responsiveness, Sustainability]*. Dhaka: Institute of Governance Studies, BRAC University.

Chowdhury, Nuimuddin, Naser Farid, and Devesh Roy. 2006. "Food Policy Liberalization in Bangladesh: How the Government and the Markets Delivered." MTID discussion paper 92. International Food Policy Research Institute (IFPRI). http://econpapers.repec.org/paper/fprmtiddp/92.htm.

CNN. 2008. "Riots, Instability Spread as Food Prices Skyrocket." *CNN.com*. www.cnn.com/2008/WORLD/americas/04/14/world.food.crisis/.

Daily Star. 2008. "Food Price Hike Major Challenge, Say Economists." *Daily Star*. www.thedailystar.net/news-detail-28283.

De Janvry, Alain, and Elisabeth Sadoulet. 2008. "The Global Food Crisis: Identification of the Vulnerable and Policy Responses." *Agricultural and Research Economics Update, Special Issue: Causes and Consequences of the Food Price Crisis*, 18–21.

GED. 2015. "National Social Security Strategy (NSSS) of Bangladesh." Dhaka: General Economics Division, Bangladesh Planning Commission, Government of the People's Republic of Bangladesh.

Gerlach, Christian. 2015. "Famine Responses in the World Food Crisis 1972–1975 and the World Food Conference of 1974." *European Review of History: Revue Européenne D'histoire*, June, 1–11. doi:10.1080/13507486.2015.1048191.

GoB. 2012. "Sixth Five Year Plan FY2011-FY2015 Accelerating Growth and Reducing Poverty Part-1: Strategic Directions and Policy Framework." Dhaka: General Economics Division, Bangladesh Planning Commission Government of the People's Republic of Bangladesh. www.unicef.org/bangladesh/Six-Five-Year-Plan-2012-Final.pdf.

Greif, Mark. 2016. *Against Everything: Essays*. New York: Pantheon Books.

Hossain, Mahabub. 2010. "The 2007–2008 Surge in Rice Prices: The Case of Bangladesh." In *Managing Food Price Inflation in South Asia*, edited by Sadiq Ahmed and Hans G. P. Jansen. Dhaka: University Press Limited.

Hossain, Naomi. 2005. *Elite Perceptions of Poverty in Bangladesh*. Dhaka: University Press Limited.

Hossain, Naomi. 2017. The Aid Lab: Understanding Bangladesh's Unexpected Success. Oxford: Oxford University Press.

Hossain, Naomi, and D. Green. 2011. *Living on a Spike*. Oxford: Oxfam.

Hossain, Naomi, and Devangana Kalita. 2014. "Moral Economy in a Global Era: The Politics of Provisions During Contemporary Food Price Spikes." *Journal of Peasant Studies* 41 (5): 815–831. doi:10.1080/03066150.2014.895328.

Hossain, Naomi, Richard King, and Alexandra Kelbert. 2013. "Squeezed: Life in a Time of Food Price Volatility, Year 1 Results." OXfam Research Reports. Oxford: Oxfam GB.

Islam, Mohammad Mozahidul. 2012. *The Politics of Food Security in Bangladesh*. University of New South Wales. http://unsworks.unsw.edu.au/fapi/datastream/unsworks:11039/SOURCE01

Karim, Lamia. 2014. "Disposable Bodies." *Anthropology Now* 6 (1):52–63. doi:10.5816/anthropologynow.6.1.0052.

Khan, Reza Ahmed, and Fatema Wadud. 2010. "Food Security and Case of Bangladesh." Report prepared by the Department of Agriculture Marketing. Dhaka: Department of Agriculture Marketing. www.foodsecurityportal.org/sites/default/files/Food%20Security%20Report%20-%20Bangladesh.pdf.

Kibria, Ashiq Mahmud Bin Gholam, Sabrina Matin, Md Motahar Hossain, and Dilara Irin. 2014. "Suspension of GSP: Image Crisis for Bangladesh." *European Journal of Business and Management* 6 (19):222–227.

Moazzem, Khondaker Golam, and Saifa Raz. 2014. "Minimum Wage in the RMG Sector of Bangladesh: Definition, Determination Method and Levels." CPD Working Paper. Dhaka: Centre for Policy Dialogue. http://cpd.org.bd/wp-content/uploads/2014/03/WP-106.pdf.

MoF. 2015. *Bangladesh Economic Review 2015*. Dhaka: Ministry of Finance of the Government of the People's Republic of Bangladesh.

Moniruzzaman, M. 2009. "Party Politics and Political Violence in Bangladesh Issues, Manifestation and Consequences." *South Asian Survey* 16 (1):81–99.

Moore, Jason W. 2014. *Cheap Food & Bad Climate*. www.researchgate.net/profile/Jason_Moore3/publication/264554597_Cheap_Food__Bad_Climate_From_Surplus_Value_to_Negative-Value_in_the_Capitalist_World-Ecology/links/53e664bd0cf25d674e9ff5dc.pdf

Motlagh, Jason, and Atish Saha. 2014. "The Ghosts of Rana Plaza: In Bangladesh, One Year After the Worst Accident in the History of the Garment Industry, Recovery Remains a Fragile Process, Justice Seems Elusive, and Reform Has a Long Way to Go." *Virginia Quarterly Review* 90 (2):44–89.

Muhammad, Anu. 2011. "Wealth and Deprivation: Ready-Made Garments Industry in Bangladesh." *Economic & Political Weekly* 46 (34):23.

Murshid, K. A. S., M. Yunus, S. M. Zulfiqar Ali, and N. Ahmed. 2009. "Re-Emergence of Food Insecurity in Bangladesh: Instability in Food Production and Prices, Nature of Food Markets, Impact and Policy." *Dhaka, Bangladesh: National Food Policy Capacity Strengthening Programme*.

NIPORT. 2009. *Bangladesh Demographic and Health Survey 2007*. Dhaka: National Institute of Population Research and Training (NIPORT).

Ortiz, Isabel, Sara Burke, Mohamed Berrada, and Hernan Cortes. 2013. "World Protests 2006–2013." *Mohamed and* CORTES, *HERNAN*, *World Protests* 2013.

Osmani, S. R. 1991. "The Food Problems of Bangladesh*." In *The Political Economy of Hunger: Volume 3: Endemic Hunger*, edited by Jean Drèze and Amartya Sen,

307–346. Oxford University Press. www.oxfordscholarship.com/view/10.1093/acprof:oso/9780198286370.001.0001/acprof-9780198286370-chapter-10

Rahaman, Mohammad Mizenur, Md Rabiul Islam, Md Abdullah, and Shah Johir Rayhan. 2011. "Female Workers' Attitudes on Industrial Disputes at RMG Sector in Bangladesh: An Empirical Analysis." *Journal of Economics and Sustainable Development* 2 (3):76–86.

Rahman, Jyoti. 2010. "Unanswered Questions about the Garments Wage Issue." *Forum*. http://archive.thedailystar.net/forum/2010/October/unanswered.htm.

Rahman, Zia, and Tom Langford. 2012. "Why Labour Unions Have Failed Bangladesh's Garment Workers." In *Labour in the Global South: Challenges and Alternatives for Workers*, edited by Sarah Mosoetsa, 87–106. Global Labour University. Geneva: ILO.

Raihan, Selim. 2009. "Impact of Food Price Rise on School Enrollment and Dropout in the Poor and Vulnerable Households in Selected Areas of Bangladesh." MPRA Paper. March. http://mpra.ub.uni-muenchen.de/37900/.

———. 2013. *The Political Economy of Food Price Policy: The Case of Bangladesh*. Helsinki: UNU/WIDER.

Ramesh, Randeep. 2008. "Bangladeshi Garment Workers Strike over Food Prices." *The Guardian*, April 15, sec. World news. www.theguardian.com/world/2008/apr/15/bangladesh.

Rashid, Mamunur, Bayazid Hasan, and Naomi Hossain. 2012. "Bangladesh: Pathways and Impacts of the Global Economic Shocks." In *Living Through Crises: How the Food, Fuel and Financial Shocks Affect the Poor*, edited by Rasmus Heltberg, Naomi Hossain, and Anna Reva. New Frontiers of Social Policy. Washington, DC: World Bank.

Rothschild, Emma. 1976. "Food Politics." *Foreign Affairs* 54 (2):285–307. doi:10.2307/20039573.

Santos, Indhira, Iffath Sharif, Hossain Zillur Rahman, and Hassan Zaman. 2011. "How Do the Poor Cope With Shocks in Bangladesh? Evidence From Survey Data." *Evidence from Survey Data (September 1, 2011). World Bank Policy Research Working Paper Series*.

Save the Children. 2009. "How the Global Food Crisis Is Hurting Children: The Impact of the Food Price Hike on a Rural Community in Northern Bangladesh."

Schneider, Mindi. 2008. "We Are Hungry!" A Summary Report of Food Riots, Government Responses, and States of Democracy in 2008.

Siddiqi, Dina M. 2017. "Before Rana Plaza: A History of Labour Organizing in Bangladesh's Garments Industry."

Sulaiman, Munshi, Monira Parveen, and Narayan Chandra Das. 2009. *Impact of the Food Price Hike on Nutritional Status of Women and Children*. Citeseer. http://citeseerx.ist.psu.edu/viewdoc/download?doi=10.1.1.541.7945&rep=rep1&type=pdf

Thorne-Lyman, Andrew L., Natalie Valpiani, Kai Sun, Richard D. Semba, Christine L. Klotz, Klaus Kraemer, Nasima Akhter, Saskia de Pee, Regina Moench-Pfanner, Mayang Sari, and Martin W. Bloem. "Household Dietary Diversity and Food Expenditures Are Closely Linked in Rural Bangladesh, Increasing the Risk of Malnutrition Due to the Financial Crisis." *The Journal of Nutrition* 140 (1):182S–188S. doi:10.3945/jn.109.110809.

Tilly, Charles. 1974. "Food Supply and Public Order in Modern Europe." In *Food Supply and Public Order in Modern Europe*, 380–455. Princeton, NJ: Princeton University Press.

The Times of India. 2008. "Food Riots Erupt near Bangladesh Capital." *The Times of India*, sec. News World Rest of World. http://timesofindia.indiatimes.com/world/rest-of-world/Food-riots-erupt-near-Bangladesh-capital/articleshow/2947769.cms

Unb. 2008. "Moeen Seeks to Dispel Fear of Silent Famine." *The Daily Star*. http://archive.thedailystar.net/newDesign/story.php?nid=28094

Woodruff, Christopher. 2014. "Enhancing Productivity in Bangladesh's Garment Sector." DEGRP Policy Essays. London: Department for International Developent-Economic and Social Research Council Growth Research Programme.

World Bank. 2008. *Poverty Assessment for Bangladesh: Creating Opportunities and Bridging the East-West Divide*. Bangladesh Development Series. Dhaka: World Bank.

———. 2010. *Food Price Increases in South Asia: National Responses and Regional Dimensions*. Washington, DC: World Bank.

———. 2013. "Bangladesh Poverty Assessment: Assessing a Decade of Progress in Reducing Poverty, 2000–2010." 31. Bangladesh Development Series Paper. Dhaka: Bangladesh World Bank Country Office.

Zhang, Xiabo, Shahidur Rashid, Kaikaus Ahmad, Valerie Mueller, Hak Lim Lee, Solomon Lemma, Saika Belal, and Akhter Ahmed. 2013. "Rising Wages in Bangladesh." 01249. *IFPRI Discussion Paper*. Washington, DC: International Food Policy Research Institute.

5 'We eat what we have, not what we want'

The policy effects of food riots and eating after the 2008 crisis in Cameroon

Lauren Sneyd

Introduction

The global media generally portrayed riots during the 2007–2008 and 2010–2011 periods simplistically, as caused by food price rises in contexts of poverty and hunger (Sneyd, Legwegoh and Fraser 2013). The main policy response across the African continent at that time was to increase staple food imports and make these 'cheap' foods more available to the urban population, which fit the presumption that the food crisis hit the urban poor hardest. Writing for *AllAfrica*, Simon Maxwell (2009) contended that the food crisis hit African cities and urban diets the hardest:

> Many Africans were hit hard in 2008 by rising food and fuel prices. With diets already stretched, and resources eroded, the numbers below the poverty line will increase, with long-term consequences. Remember, hunger is corrosive of long-term life chances: it undermines the ability to grow, to learn, to stay healthy and to earn. . . . The first to suffer this time will not be rural people, far from the capital city, losing crops and animals, and often children, to the lack of rain. Instead, jobs will be lost in towns, amongst the vocal and educated. There were food riots in Africa last year. . . . And tension is rising.

What were reported as the food riots of 2008 were actually more complex and significant than riot events, both in their origins and in their impacts on urban hunger (Hossain and Green 2011; Mason *et al.* 2011; Sneyd *et al.* 2013). The African media tended to portray the food riots as being caused by a more complex set of factors, including citizen dissatisfaction and people's ability to mobilize (Sneyd *et al.* 2013). And some researchers argue that the protests across Africa at that time were not food riots at all (Branch and Mampilly 2015). So simply throwing more 'cheap' food at the problem would not touch the simmering, underlying problems that sparked the protests in the first place.

An emergent body of work on the food riots of 2007–2008 suggest that the riots were linked to food inaccessibility on the one hand and inequality, poverty and the need for better governance and greater equality on the other (Dia Kamgnia 2011; Sneyd *et al.* 2013; Berazneva and Lee 2013; Dimova and Gbakou 2013; Hossain *et al.* 2014). This is contrary to a commonly held assumption about the causes of food riots. For instance, the authors of the Foresight Report (2011) assume that a food price rise triggers riots regardless of situation and context, and that it is necessary to produce more food to avoid riots in the future. But the reality is more complex: the food price crises have made accessing adequate food and maintaining nutritious diets more difficult for urban dwellers and connects them more deeply to a globalized food system (Pinstrup-Andersen 2010; Legwegoh *et al.* 2015; Clapp and Scrinis 2016). During the crisis, it seemed that it was easier to fill people's bellies than tackle a more complex set of factors such as disenfranchisement, marginalization and accessibility.

In Cameroon, the distinction was no different. Drawing on interviews in markets and households in three cities in Cameroon, it was learned that the rice subsidies introduced during the crisis changed people's diets and the local forest food economy in negative ways, replacing traditionally diverse diets with refined white rice-based dishes. The reliance on imported rice creates a vulnerable urban population, especially when world prices for rice remained volatile after 2008 (Moseley, Carney and Becker 2010). The reality of relying on the world market for rice has had a considerable impact on urban dwellers realizing food security and has set the stage for a nutrition transition in this Central African country (L. Sneyd 2013, 2016b).

Methodology and approach

Two questions guided this study: What was the impact of the 2008 food riots in Cameroon on food security? What was the long-term impact of the food riots and rice subsidy on people's diets after 2008? To answer her research questions, the author made five field visits to Cameroon from 2010–2013 to conduct urban food security research and to ask questions about the 2008 food riots and impacts on local food economies after the price spikes.[1] During the field visits, over 100 elite interviews were conducted with officials from the Food and Agriculture Organization (FAO), World Food Programme (WFP), the World Bank and the International Monetary Fund and relevant government officials and local researchers. In addition to the in-country fieldwork, the study employed mixed methods, including a review and analysis of media sources, reports and academic studies. The author also hired a team of 12 local researchers to conduct qualitative interviews with urban food traders and their customers on food security, dietary change and the 2008 unrest. These interviews were primarily managed in urban food markets during the wet and dry seasons in 2012–2013. In total, 409 people were interviewed for the study. The data collection happened in two phases. In the first phase, 371 interviews were completed; and in phase two, 38 in depth-interviews were completed. The two phases covered 24 markets in three cities – the French-speaking capital and smaller English-speaking cities. See Table 5.1 for markets and number of interviews from the first phase.

Table 5.1 Market and number of interviews in Cameroon

Country and City	*Market Name*	*# Interviews (n = 371)*
Southwest: Buea	Soppo/Buea	17
	Muea	11
	Ekona	6
	Mile 4	13
Muyuka	Muyuka (formerly Soppo)	16
Mutengene	Mutengene	19
	Mutengene 2	8
Limbe	Limbe	32
	Likumba	12
Centre: Yaoundé	Mokolo	54
	Mvog Mbi	37
	Nkol Eton	19
	Mvog-Betsi	25
	Elig-Edjon	29
	Essos	37
	Biyem-Assi	11
	Huitieme	25

The multi-focal approach was adopted to capture 'sounds from the ground' or impressions from people who may have been involved in the unrest, to learn more of the lived experience of the unrest and impacts on household food security. Interview questions ranged from questions about household size, employment and family assets to questions about household food shopping, what happens when food prices are high and what the households does when there is not enough money to buy food. Also included were questions about the foods that are commonly consumed when food prices are high and questions about the cheapest meal consumed in the household and the most favorite meal. Questions were also asked about what was remembered about the riots of 2008, what foods do you need to feel food secure and happy, and how has your diet changed over time. The quotations from individuals from the markets are not outliers but are representative of the theme under discussion. Accompanying the quotes from the surveys is the market name and the date the quotation was recorded. The author has not changed the quotes, relying on the informants' recorded account.

Table 5.2 characterizes the data by age, gender and reported class based on employment, income and assets. The data predominately represents middle class (65 per cent) females (86 per cent) between the ages of 20–40 (54 per cent). The data depicts a variety of occupations, primarily food vendors (~171) on precarious incomes in the informal sector. However, household interviews were also conducted with house cleaners (26), business owners and employees (17) teachers (18), civil servants (12), housewives (12), nurses (7), secretaries (7), students (8), lawyers (4), retirees (4), journalists (2), professors (2), aestheticians (2), hotel industry employees (2), a doctor, a naturopath, a landscaper, a builder, a prison guard, a taxi driver, an accountant, a freelance writer and a decorator. The

Table 5.2 Characterization of interview data by age, gender, class

Interviews (n = 367)	*Total #*	%	*Interviews (n = 371)*	*Total #*	%	*Households (n = 197)*	*Total #*	%
Age < 20	3	0.8	*Male*	54	14.5	*Poor*	44	22.6
Age 20–40	197	53.7	*Female*	317	85.5	*Middle*	126	64.6
Age 40+	135	36.8				*Rich*	16	8.2
No answer	32	8.7				*No answer*	9	4.6

participants represent a middle class in Cameroon from popular neighbourhoods, where contingent employment, employment in the informal sector, short-term employment as civil servants or seasonal workers on plantations are the norm (Bernards 2015; L. Sneyd 2015).

The majority of urban dwellers in Cameroon spend more than half of their income on food (from 50–75 per cent). This does not leave much room in the budget for other basic needs (Sneyd 2016a; 2016b). Households in the highest income group spend 25 per cent of their income on food. By contrast, those in the lowest income group spend 66 per cent of their income on food. The households interviewed mentioned various coping strategies they adopt when food accessibility challenges arise. They eat fewer meals, a smaller quantity of food and lower quality food, and consume less diverse diets. When factoring in dependents and household size this study includes over 2,111 individuals. The author makes no claim from this case study that data are representative of the wider populations of these cities or quantitative rates of food insecurity in these cities. Instead, the data represents a snapshot into the lived and daily realities of food provisioning for residents of Yaoundé, Limbe and Buea.[2]

The data from the interviews were organized using dimensions established by Olivier de Schutter, the former UN special Rapporteur on the Right to Food, to better understand the food security situation in Cameroon (see also A. Sneyd, Legwegoh and L. Sneyd [2015]). De Schutter visited Cameroon in 2012 and his report on the mission focused on the widely accepted dimensions of the right to food, such as: the availability of food, the accessibility of food and the adequacy of food (de Schutter 2012; A. Sneyd, Legwegoh and S. Sneyd 2015; Clapp 2017). This chapter focuses on these dimensions to better understand the impact of the unrest on food security and the subsequent rolling out of social safety nets, such as the "universal subsidies, import and value added tax (VAT) exemptions" that were used by the government to mitigate the negative effects of high food, energy, and transport costs (del Ninno and Tamiru 2012: xiv).

The chapter proceeds first by detailing the events of the 2008 food riot in Cameroon and the policy responses. Second, it turns to three sections that detail the three dimensions of food security (availability, accessibility and adequacy) and illustrate with interview data how each of the dimensions were impacted by the unrest. The paper then enters into a discussion of the experience of the accessibility/adequacy tradeoff amongst urban dwellers. The chapter concludes on a somewhat reflective note that re-examines questions about food riots in the scholarly literature then delves back into the Cameroon case and characterizes the dietary change that helped the urban population cope with the food price crisis.

Cameroon's 2008 food riots

Resentment against the ruling elite has simmered in the country for decades as the majority of the population and the youth in particular experience insecurity, unemployment and poor living conditions. According to Cameroon's National Human Rights Observer, the social and economic situation:

> has continuously been deteriorating, leading to an unbelievably worsened situation especially in the areas of corruption, unemployment, poverty, misery, insecurity. The disruption of water and electricity supply has become too frequent. Health and education have become less accessible. The living conditions of students have deteriorated: no scholarships, school fees steadily rising and insanitary condition of university hostels . . . most Cameroonians feel that the government and her sclerotic members of the ruling class are completely indifferent to their problems.
>
> (2008: 8)

The events of 2008 in Cameroon reported globally as 'food riots' were motivated by a mix of political concerns: "the draft amendment of the Constitution, which was seen as an extension of the mandate of President Paul Biya and his regime of bad governance and the rise in price of fuel and food" (National Human Rights Observer 2008: 5). Together, these factors combined to trigger the popular uprising. The protesters voiced their demands for the realization of their socio-economic, civic and political rights. The protests were located in major cities of five regions in the southern parts of Cameroon, namely: the Southwest, Northwest, West, Centre and Littoral. Figure 5.1 shows the cities where the strikes and protests were located and the markets in Yaoundé, Buea and Limbe where the interviews were conducted for this study.

Throughout 2007, several top officials made repeated calls for the constitution to be amended to allow the Cameroon People's Democratic Party (CDPM) "to seek a third mandate in 2011 against the provisions of the constitution" (Agou 2011; National Human Rights Observer 2008: 7). On 31 December 2007, on the occasion of his end of year speech, President Paul Biya and his ruling party publicly announced that he would reform the constitution to extend his then-26 years in power to seek a third mandate in the 2011 election cycle, thus paving the way for the president to become head of state for life[3] (Amin 2012; Observatoire National des Droits de L'homme 2009 and 2011; Friends of Cameroon 2008; Reuters 2008c).

Sensing the trouble that a constitutional change could bring about, the governor of the Littoral region issued an order on 15 January 2008 prohibiting

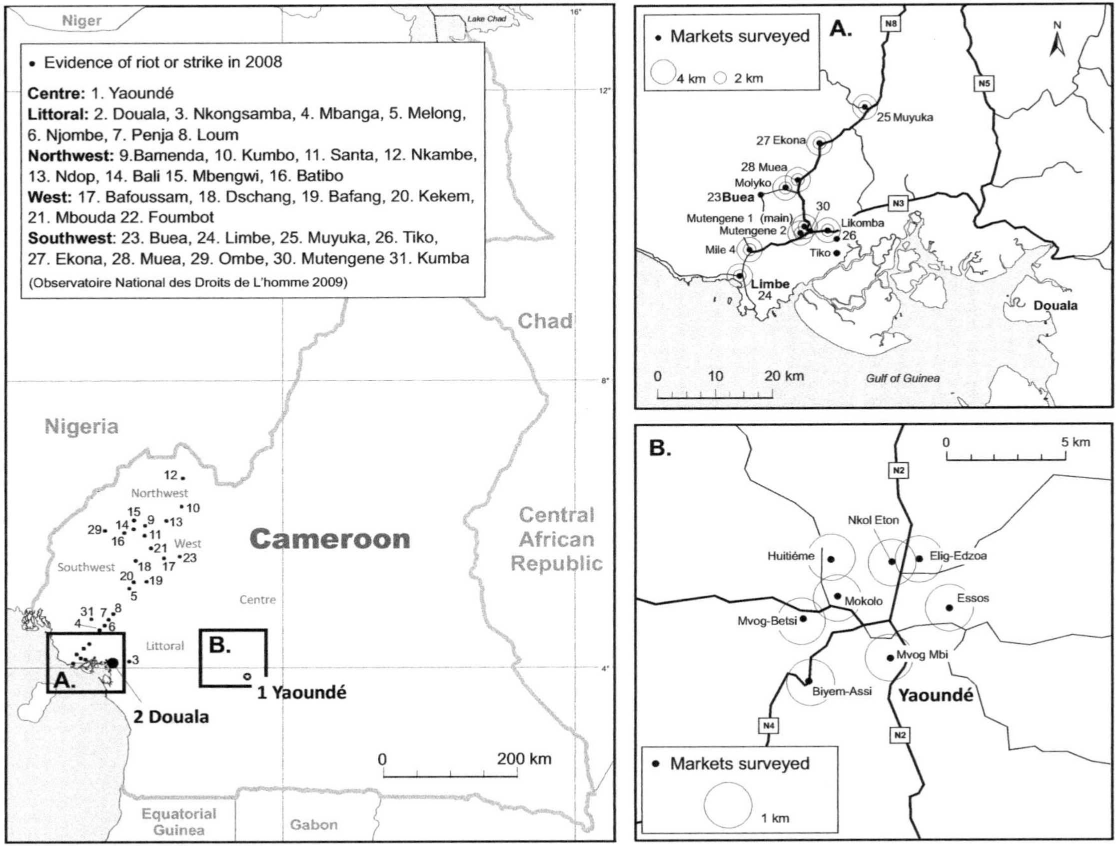

Figure 5.1 Cameroon food riot map

demonstrations and other unauthorized occupations on public roads for an indefinite period of time (Observatoire National des Droits de L'homme 2009). The order was largely unsuccessful as strikes unfolded in cities across southern Cameroon. On 23 February, TV and radio stations in Douala that would have covered criticisms and debate of the proposed constitutional amendment were shut down. Protests were already organized in January by transportation unions to protest the high cost of fuel; however, given the new constitutional reality, the organized protests took on greater meaning for protesters. One trader said, "we strike when we are not happy with certain decision taken by the government,"[4] and another participant in the study mentioned that a common reason to protest is "when we are sick and tired of certain things. We express our miscontent [sic] to the government authorities to deal with any problems we may have."[5] Cameroon's Human Rights Observer documented the unexpected response:

> Many youths spontaneously descended to the streets to express their disillusionment and loss of hope for a better future. The strike then became a widespread movement. No political organization or trade union instigated the protestors. It was all spontaneous. The cities were totally paralyzed. Peaceful demonstrations could be seen everywhere. Rallies were held at junctions of main streets.
>
> (2008: 10)

From 23–28 February, shops, bakeries, pharmacies, hotels, restaurants and filling stations were ransacked and looted. Government and perceived government buildings and offices were targeted. The protest slogans recorded below show the mix of political and food price demands:

> They were in the streets in the thousands in different cities demonstrating and marching, and while a recurring theme was Biya Must Go (*Biya doit partir*) other placards were equally revealing: *Popol tu seras pendu avec ta constitution; on a faim, non à la vie chère; trop c'est trop*; Do Not Touch The Constitution; Halt The Cost Of Living; We Need Jobs; Paul Biya, *La jeunesse avant la constitution*; Hungry Man Is Angry Man; No More Biya; No Constitutional Amendment; Down With France; Biya Is Old And Tired; Constitutional Change Only Over Our Dead Bodies; No Life President; End the Corruption; *Non à la vie chère et à la vie clochardisation de Camerounais; Baisez le prix du tapioca sur le marché; Non à la fermeture des radios et télevisions privies.*
>
> (Amin 2012: 28)

It quickly became the first protest in the nation's history in which "minute- by-minute events were covered by social media" (Amin 2012: 32). By 27 February, a strike had taken place in 31 cities, including Yaoundé, Douala, Bamenda, Bafoussam, Buea, Limbe, Tiko, Muea, Mutengene and Kumba (see Figure 5.1).

It was a time when "there were a lot of policemen on the street," as police were mentioned 60 times in the interview data. One woman in Buea remarked:

> It was not a very good time for us. I almost got shot by a police officer in Mutengene. I was around an area where young boys were throwing stones at the police officers and one of them almost shot me.[6]

Reactions from the police, including mass arrests, live bullets, tear gas and barricades occurred in all protesting cities throughout the country.

On 28 February, Biya issued a "stern warning to Cameroonians to stop the mayhem, vowing to prosecute to the fullest those 'demons,' and 'apprentice sorcerers,' who had 'manipulated' Cameroon's youths" (Amin 2012: 33). After the warning was broadcasted in Yaoundé, an informant mentioned that fires and smoke could be seen across the city. Protesters were dismayed:

> Nothing on the issue of constitutional reform, the high cost of living, rising prices and chronic unemployment of young people was addressed. The disappointment was widespread among the population. The demonstrators were thereafter regarded and treated as enemies of the state.
>
> (National Human Rights Observer 2008: 15)

To combat the protesters, Cameroon's special military units such as the *Bataillon d'Intervention Rapide* (BIR), an elite paramilitary force, was transferred to Yaoundé, Douala and other cities in the country. The BIR was created in 2001 to combat armed groups and gangs along the nation's borders with Chad, Nigeria and the Central African Republic, and has been used since to fight Boko Haram in the North (Amin 2012; Observatoire National des Droits de L'homme 2011; IRIN 2008; Hilsum 2015). Joining police forces, the BIR fired live bullets, tear gas, stun grenades and water hoses indiscriminately into protesting crowds. One trader explained:

> We felt unsafe in this same market and we ran for safety. Can't remember the cause, but the military came and people started running, boys burnt [things] and scattered. The young men were angry, some were arrested some were beaten up.[7]

The response resulted in both "financial losses and loss of human life."[8]

The reaction of the government resulted in several violations of human rights including: excessive force against an unarmed population; targeted arrests and arbitrary detentions; torture and cruel and degrading treatment; violations of the rights to freedom of expression, opinion and information; and arbitrary executions (Observatoire National des Droits de L'homme 2009, 2011). During this time, the death toll has been debated. The government estimates around 40 deaths, whereas civil society groups in Cameroon suggest the number is well over 139. Also disputed is the number of arrests of as many as 3,000 individuals (Observatoire National des Droits de L'homme 2009, 2011: 23).

President Biya initiated a program to subsidize petrol, and to suspend import duties on fish, rice and cooking oil to lower their prices[9] (Reuters 2008a, 2008b). It has been documented that even though the president "moved to reduce prices on food and petroleum products, the protests didn't stop" (Amin 2012: 22). One 70-year-old woman remembered an old proverb: "a hungry belly has no ears (*un ventre affame n'a point d'oreille*), it is impossible to talk to someone torn by hunger."[10]

The subsidies only applied to imported food and locally produced and forest food remained expensive (L. Sneyd 2016a). The massive presence of military and police brought calm to the country on 29 February, although fear was felt by many residents. Over time, the government's policy seems to have appeased the urban population in the short term, but some argue – and the World Bank concluded – that another price spike could drain domestic savings and disrupt balance of payments (Galtier and Vindel 2013). Not only of concern to the economy, informant responses in this study demonstrate the challenges facing the urban poor in the context of overwhelming food accessibility constraints. The following sections explore in greater detail the food security situation after the 2008 riots.

The impacts on the right to food: Issues of availability, accessibility and adequacy

Availability issues

The simultaneous rise in fuel prices compounded the local effects of the food price rise. Not only was food more expensive, it was more expensive to move it around the country. In the Cameroon case, food was available in 2008, but for many, not accessible. In the Cameroonian cities surveyed, the strikes were remembered as a time of low stocks in the markets and food shortages at home. The scarcity was linked to instability and violence, and the closure of markets. There was also an absence of supplies due to fires, burning of market structures and food and the destruction of perishable products. Tires and other materials were burned on highways so the entrances into towns were impassable.

The actual ways in which people responded to the crisis varied in Cameroon. The transporters strike and the subsequent halt in the movement of food and goods had the greatest impact on food availability. When the transportation networks were stalled, artificial scarcity was created, fueling further unrest and driving food prices higher. During this time, the rise in the cost of transport prevented most traders from buying wholesale food to sell because "vehicles were not bringing food to the market."[11] According to the traders, this strategy "paralyses the market and the food supply chain is altered."[12]

In Yaoundé, market officials "closed food markets until the calm returned."[13] The consequences of riots and the closure of markets were deeply felt: "Consumers are forced to go without their products and forces people to suffer as food is hard to find." Affected traders tried to temporarily sell in other markets until their market re-opened: "We don't sell anymore. I tried to find a place in *Mvog Mbi* but they chased me away (*on ma chassé*)."[14] One woman said:

> If there is a riot and the market is closed I won't have anything to eat. We run around trying to find food but cannot. We would be obliged to get our products elsewhere which incurs added expenses.[15]

Because of the violence, traders and officials closed markets to maintain calm (see Table 5.3 for a list of code counts from the interview data). Traders mentioned how "markets stayed closed for a week which penalized both the vendors and their clients."[16] When asked what the consequences are of riots and the closure of the market on your activity/sales, many traders state: "We don't sell. When the market is closed I don't work. If there is a riot, I don't go to sell there is lower sales and it slows down activity."[17] The consequences of riots and the closure of markets are deeply felt: "consumers are forced to go without their products and forces people to suffer as food is hard to find."

Accessibility issues

An International Food Policy Research Institute (IFPRI) paper calculated the change in food prices from June 2007 to June 2008 in food markets across Africa,

Table 5.3 Cameroon's 2008 protests: Themes and code frequency from market interview data

Event	*Availability*	*Accessibility*	*Adequacy*
Riot 215	Shortage 165	Buy 1,425	Rice 1,016
Strike 98	Market closure 104	Money 425	Plantain 1,010
Police 60	Closed markets 61	Less expensive 122	Less 799
Government 50	Losses 38	Food reserves 61	Expensive 518
Taxi 46	No sales 37	Forced (to eat) 54	Corn 414
Death 26	Douala 36	Manage 53	Import 330
Drivers 23	We don't sell 30	Borrow 46	Cocoyam 279
Transportation 22	Scared 23	Difficulties Provisioning 44	Change 224
Injuries 17	Stops 16	Make do 36	Cheap 172
Intervention by armed Forces 15	Yaoundé 14	Credit 13	Small 147
Crisis 12	Road 10	Cannot provision 13	Without 140
Violence 8	Prevents 8	Not accessible 9	Diet 139
Fire 7	Blocked 7	Spend too much 8	Cassava 126
Flee 7	Disturbed 6		Reduce 109
Confrontations 5	Paralyzed 4		Local 109
Citizens 3			Okra 93
Military 3			Rice subsidy helps 62
Protest 0			Famine 60
			Diversity 57 (describing market choice)
			Insecurity 33
			Poor quality 11
			Reduce quantity 15
			Dangerous (food) 5

including Cameroon (Minot 2011). The calculation from Yaoundé revealed that the retail prices for many food items increased substantially: cassava (16 per cent), maize (29 per cent), plantains (14 per cent), red beans (17 per cent), rice (36 per cent) and wheat flour (51 per cent) (Minot 2011: 15). When food prices rise, participants mentioned, "I buy less or same quantities, sometimes I don't buy at all. I mostly reduce quantity and at times go for the cheap ones."[18]

Urban Cameroonians spend 50–75 per cent of their income on food, yet "the food basket of each household is becoming more and more meager because life is expensive in our country in general and in the city of Yaoundé in particular."[19]. In Yaoundé, "food products are very expensive, to eat well we are forced to spend too much and this is an issue for impoverished families."[20]

'Forcing', 'forced to change' and 'forced to spend' was a major theme from the interview data. Another individual explained in more detail:

> [T]he riots really disturbed a lot of people. Everybody was complaining about rise in prices, especially sugar, flour and rice. It was such a bad experience. But I think the situation is the same now although we are now used to food prices being high.[21]

Households in Limbe, a coastal city outside of Douala, remembered "a limited quantity of food was brought to the market since buses were not moving and drivers were not working because of high prices in fuels."[22] One customer in Buea remarked:

> [W]hat I remember is that it made prices increase more. We bought very little food for so much money because cars were not allowed to run; the roads were blocked. When the market was ablaze most of the business men and women lose most of their items.[23]

When households do not have enough food or enough money to buy food, the most commonly named strategies include:

- *I manage with the little I have*
- *I am stricter about how much food I serve and other than the children we only eat one meal a day*
- *I make a cheap meal with rice and palm oil or garri (cassava) and water*
- *I reduce the quantity of food I eat*
- *I eat one or two meals a day*
- *I harvest from my farm*
- *I borrow from family, savings group, or merchant*

From these data, it can also be learned that eating fewer meals also means eating less at meals, eating lower quality foods and eating cheaper options (see also Table 5.3). Responses from the survey show that 54 per cent of the individuals surveyed consumed only two meals a day and if these meals already include dishes that are of low quality and diversity, then nutrition is negatively impacted (L. Sneyd 2016b). When prices for foodstuffs are elevated, women mentioned buying in smaller quantities, changing their menu, and making food every two days, for the following two days, based on what is affordable. The food security situation in most households interviewed is tenuous at best.

Adequacy issues

The government continued to subsidize rice after 2012, making it available, accessible and cheap. Since the riot and subsequent government subsidies, many people say that imported food, especially the rice subsidy "helps us survive." It *helps* was an overarching theme (see also Table 5.3). The rice that is imported is the milled white rice, which is devoid of the husk and bran. This often results in a version of rice that is quick to cook, but is low in fiber and vitamins. Cheap rice shifts purchasing patterns of urban households away from traditional, local foods (such as cassava, cocoyam, plantain, local rice and wild vegetables), which are nutritionally more beneficial and Cameroon has a comparative advantage in producing (L. Sneyd 2013). Older women surveyed were shocked by the frequency at which their younger counterparts prepare and eat rice. To these older women, rice was a special occasion dish, often being served for the Sunday meal or for holidays with chicken (L. Sneyd 2013). When asked how many times they prepare rice per week, most women stated that they serve rice three to four times, some even claiming "in my household we eat rice five times a week."[24] The role of rice in the alimentation of households was most commonly stated as being "primordial" as "it allows me to make ends meet in difficult times." Other women used adjectives such as "important" and "profound." Almost all households interviewed stated that rice serves "a big place in the household menu, because it is not very expensive, it is accessible and easy to cook, it is the most consumed product in our household." The origin of the rice most commonly consumed is from Thailand, while the heartier Cameroonian rice is reserved for holidays.

The subsidies excluded locally produced food and prevented people from buying local food and local forest food, which continue to experience high prices (L. Sneyd 2013, 2016b). During the interviews, participants were asked to list the foods they need to feel food secure. The most commonly named were: rice, plantain, groundnut, couscous, palm oil, fish, tomato, fruits and vegetables. One respondent said: "I do not have food security; I do not eat fresh meat and fish as I would like." Many people said that local food in Yaoundé is too expensive. When asked which imported food products are more available to you than local products, the greatest frequency in the response was rice, pasta, fish, and sardines. Households in Yaoundé described these products as being "available to everyone, they are less expensive than local products." Other comments included: "We need to learn to eat our own products and the Cameroonian state must review the food prices" and "Food in Yaoundé is expensive, to eat well we are forced to spend too much. The government should watch and review the prices." A not-so-surprising finding from interviews is that Cameroonians want to eat foods from Cameroon.

It was perceived that imported rice is bad for the economy as it prevents people from buying local. Others think that the imported foods are of poor quality, they are toxic, full of fertilizers and taste is bad. They can "make you sick and gain weight." Many participants even said rice is "a good thing for the poor classes. The foods coming from outside help to reduce hunger, especially rice." While rice may be helping accessibility challenges, it is not improving the adequacy of diets. Overwhelmingly participants indicated that "our diet changed."[25]

The accessibility/adequacy tradeoff

In cities in Cameroon the food price crisis has had a strong impact on households' food security. People reported eating less and changing their diets. During the interviews, participants were asked to name and explain the ingredients for and technique of preparing meals in the household; the cost, who taught you to cook this meal and how often do you eat it? The questions focused on four different categories: the meal that is prepared the most in the household, the favourite meal, the most expensive meal and the cheapest meal consumed in the household. The data from the interviews yielded 420 recipes. What was learned from these series of questions and the recipes was that rice is not an accompaniment to the dishes prepared with food products from southern Cameroon's humid forest zone (L. Sneyd 2015). Dishes made with ingredients from the Congo Basin's forest emerged from a different culinary history than dishes that are traditionally prepared with rice (L. Sneyd 2013). Traditional dishes made with local forest food products that make up the routine cuisine accompany cassava or plantain, such as *eru* (a stew made with Gnetum leaves from wild vines, cow skin, crayfish and red palm oil), *ndole* (a hearty stew of leafy greens, raw groundnuts and beef) and *mbongo* (a dark stew made with charred wild spices and a protein, usually fish). It is culturally inappropriate and unpalatable for the routine dishes to simply be served with rice. Temporally, this is important, as four years after the food riots people said that nothing has changed save for a *decrease* in their dietary diversity: "We eat what we have, not what we want."[26]

Refined white rice helped facilitate a dietary change; however, it can be argued that frequent consumption of conventionally-produced industrial and imported rice should be a coping strategy and not a long-term solution. The policy responses – importing and subsidizing rice – have made diets less 'sustainable,' and undercut the adequacy of diets through making healthier, more sustainably-produced local foods, whether they were forest foods or less accessible agricultural products. Subsidizing rice to appease the rioting urban population further marginalized the poor from achieving the right to food and maintaining acceptable levels of household food security. The tradeoff here meant that due to accessibility challenges, the adequacy of diets suffer as households were forced to change their eating habits, eat a less diverse diet and eat less frequently.

Returning to the data, it was not only increased refined white rice consumption that changed diets in cities, but the accompaniments to the starches changed as the local vegetables, meats and spices were less accessible; prices for those foods remained high into the present (L. Sneyd 2013; Rao 2014). The accessibility/adequacy tradeoff is the reality for many in urban Cameroon. Circumstances dictate that urban residents must make a compromise for cheaper, less diverse food options instead of maintaining a diet of diverse proteins, starches and vegetables. This was demonstrated in the data. Participants from cities in Cameroon mentioned eating rice three to five times a week. Hence, low-cost subsidized rice plus rising food prices not only priced more diverse local food options out of the reach of most, but also enabled a situation where there was a reduction and substitution of diverse local foods for most city dwellers. This led urban residents to substitute cheaper, less nutritious and less diverse foods – in Cameroon's case, conventionally-produced, industrial, refined white rice, imported fish and imported tomatoes – for healthier, more diverse and locally produced foods. Data collected on dietary change showed that a new cheap dish is emerging across the urban landscape in Cameroon that puts these items into the same pot and that dish was consumed three to five times a week. The resulting dietary and culinary shift marks a turn away from the routine cuisine of a local starch (either cassava, cocoyam or banana plantain) prepared with a saucy meaty stew (made with beef, local fish and crayfish, green leafy vegetables and spices from the surrounding forest) toward a new subsidized, refined and cheap white rice dish. In the short term, the coping strategy of increasing rice consumption helps people cope with food accessibility challenges but it does not fix or address their challenges in the longer-term; it inadvertently ties them to a globalized food system relying on trade and un-nutritious and homogenized foods.

Discussion and conclusions

The evidence showed that political disenfranchisement, anger and outrage over an unjust political situation enabled the conditions for the riots and riots cannot solely be understood to be desperation over the rising cost of food. This also relates to Hossain and Kalita's conclusion on the political possibilities of action: "not all hungry people riot, and for many collective action is a risk" (2014: 14). This finding is also in line with Demarest's (2015) argument that there are numerous triggers that can spark a riot event, and rising food prices alone cannot explain this relationship. The Cameroon case raises questions about riots more generally: is it a food riot if the price spike coincides with the government moving to change the constitution and extend presidential terms? Is it a food riot or something different (Pritchard and Pakes 2014)? Can it still be called a food riot when the protests become a national movement that results in large-scale human rights violations?

The grievances are diverse yet each is intimately connected to civil and political rights, human rights, sustainability and the right to food. The longer-term reality in Cameroon raises additional questions: was the regressive subsidy scheme a solution or a last-ditch attempt at political legitimacy? Was the subsidy a crude but effective resolution for a government in a globalizing food system – when a state was losing its legitimacy? Is it acceptable to fill people's bellies without focusing on dietary diversity? Why is subsidizing and supporting domestic agriculture not considered as a way out of a food crisis (Lappé *et al.* 2013)? What about provisions to support and improve rural productivity of small holder units that will contribute to the achievement of the UN Sustainable Development Goals, or transformation for gender and food security in Cameroon's forest food trade (L. Sneyd 2016a)? These questions may not yield easy answers, but it is clear that the government response, by making cheaper rice available, further accelerated the nutrition transition in Cameroon (Steyn and McHiza 2014). The long-term effects have weakened people's food security because they are now more dependent on volatile global markets for food of less dietary diversity. The food crisis in Cameroon changed diets towards more monocropped foodgrains, and – for the urban population, at least – a dependence on imports. The Cameroon case demonstrates the challenge of making foods available that people want to eat and that provide a diversity of nutrients and vitamins versus simply filling bellies.

Notes

1 The Social Science Research Council of Canada (SSHRC) funded this research. The author wishes to thank the following individuals for their various contributions to the project: Evan D.G. Fraser, Craig Johnson, Bruce Frayne, Jonathan Crush, Olufunso Somorin, Florence Munoh, Phil René Oyono, Stella Asaha, Akoko Guy Martin Maurice, Serge Kombo Samba, Ngu Arrey Monica, Caleb Yengo Tata, Roland Ndah Njoh, Eugene Loh Chia, Fomeni Anne Christelle, Ashu Augustine Ayuk, Abdou Njida, Ekane Ivo, Mary Nana Nibba, Hubert (Nana), Patrick Wight, Brendan Schwartz, Naomi Hossain, Patta Scott-Villiers, Adam Sneyd and Marie Puddister for assistance with the map, and the hundreds of individuals who shared a glimpse into their lives after the food price crisis in Cameroon.

2 Yaoundé is the French-speaking capital and the base of operations for many multilateral organizations for the region, including various UN agencies, the World Bank and the International Monetary Fund. The Banque des Etats de l'Afrique Centrale (CEMAC) is located in Yaoundé and operates for the CEMAC region (including Chad, Central African Republic, Equatorial Guinea, Gabon and the Republic of the Congo). Yaoundé is also the hub for numerous INGOs including Greenpeace and the World Wildlife Fund, and domestic NGOs. Limbe was an old English colonial town, Victoria. A major oil refinery is located off the coast of the city in the Gulf of Guinea. Cameroon Development Corporation (CDC) – an agro-industrial complex that grows, processes and exports many tropical crops such as tea, bananas, palm oil and rubber – is also based out of Limbe and a major plantation employer for the region. At the foot of Mount Cameroon and up the road from Limbe is Buea, an English-speaking university town and the capital of the southwest region and an electoral district of the Social Democratic Front (SDF), a prominent English-speaking opposition party. Cameroonian Pidgin English is spoken in both Limbe and Buea.

3 Biya had succeeded Ahidjo in November 1982 and became with Eduardo dos Santos of Angola and Robert Mugabe of Zimbabwe, "one of the longest-reigning presidents in Africa" (Albaugh, 2011). Fru Doh (2008) documents the history of growing elite power and political favoritism in his book suitably titled: *Africa's Political Wastelands: The Bastardization of Cameroon*. In the past, this system has been described as neopatrimonial, but new dynamics of capitalism in Cameroon have recently been identified (see A. Sneyd 2014).

4 Interview with author in Mokolo, July 2012.
5 Interview with author in Mokolo, July 2012.
6 Interview with author in Likomba, July 2012.
7 Interview with author in Mutegene, July 2012.
8 Interview with author in Mvog-Betsi, July 2012.
9 According to a World Bank Report, Cameroon's social safety nets are ad hoc at best and regressive. The government "subsidized the price of several food items including maize, flour, fish, rice, and wheat to the tune of 60 billion FCFA [about USD $104 million] on average annually between 2007 and 2009 (an average of around 2.42 percent of the government budget)" (del Ninno and Tamiru 2012: xiv).
10 Interview with author in Mokolo, November 2012.
11 Interview with author in Likumba, July 2012.
12 Interview with author in Nkol Eton, July 2012.
13 Interview with author in Mokolo, November 2012.
14 Interview with author in Mokolo, July 2012.
15 Interview with author in Mvog Mbi, July 2012.
16 Interview with author in Essos, November 2012.
17 Interview with author in Mokolo, November 2012.
18 Interview with author in Likomba, July 2012.
19 Interview with author in Yaoundé, November 2012.
20 Interview with author in Mvog Mbi, November 2012.
21 Interview with author in Mile 4, July 2012.
22 Interview with author in Limbe, July 2012.
23 Interview with author in Mutegene, July 2012.
24 Interview with author in Mokolo, November 2012.
25 Interview with author in Elig-Edjon, November 2012.
26 Interview with author in Tiko, November 2012.

References

Agou, Melvin. 2011. "Cameroon's Presidential Election: Will the Votes Count?" *Brookings Opinion*. 6 October 2011. www.brookings.edu/opinions/cameroons-presidential-election-will-the-votes-count/

Albaugh, Ericka A. 2011. "An Autocrat's Toolkit: Adaptation and Manipulation in 'democratic' Cameroon." *Democratization* 18 (2):388–414.

Amin, Julius A. 2012. "Understanding the Protest of February 2008 in Cameroon." *Africa Today* 58 (4):20–43.

Berazneva, Julia, and David R. Lee. 2013. "Explaining the African Food Riots of 2007–2008: An Empirical Analysis." *Food Policy* 39:28–39.

Bernards, Nick. 2015. "The Internationalization of Labour Politics in Africa." *Critical African Studies* 7 (1):7–25.

Branch, Adam, and Zachariah Mampilly. 2015. *Africa Uprising: Popular Protest and Political Change*. London: Zed Books.

Clapp, Jennifer. 2017. "Food Self-Sufficiency: Making Sense of It, and When It Makes Sense." *Food Policy* 66:88–96.

Clapp, Jennifer, and Gyorgy Scrinis. 2016. "Big Food, Nutritionism, and Corporate Power." *Globalizations* 14 (4): 578–595.

Demarest, Leila. 2015. "Food Price Rises and Political Instability: Problematizing a Complex Relationship." *European Journal of Development Research* 27:650–671.

de Schutter, Olivier. 2012. "Report of the Special Rapporteur on the Right to Food: Mission to Cameroon." United Nations General Assembly A/HRC/22/50/Add.2.

del Ninno, Carlo, and Kaleb Tamiru. 2012. *Cameroon Social Safety Nets: Africa Social Safety Net and Social Protection Assessment Series*. Washington, DC: World Bank.

Dia Kamgnia, Bernadette. 2011. "Political Economy of Recent Global Food Price Shocks: Gainers, Losers and Compensatory Mechanism." *Journal of African Economies* 20(suppl1):i142–i210.

Dimova, Ralitza, and Monnet Gbakou. 2013. "The Global Food Crisis: Disaster, Opportunity or Non-Event? Household Level Evidence From Coté d'Ivoire." *World Development* 46:185–196.

"Foresight: The Future of Food and Farming." 2011. Executive Summary. The Government Office for Science, London. www.bis.gov.uk/assets/ bispartners/foresight/docs/food-and-farming/11–547-future-of-food-and- farming-summary.pdf

Friends of Cameroon. 2008. "Negotiations Sought After Deadly Riot." 29 February 2008. www.friendsofcameroon.org/2008/02/29/breaking-news-anti-government-riots-hit-cameroon-capital/

Fru Doh, Emmanuel. 2008. *Africa's Political Wastelands: The Bastardization of Cameroon*. Bamenda: Langaa RPCIG.

Galtier, Franck, and Vindel Bruno. 2013. "Managing food price instability in developing countries: a critical analysis of strategies and instruments." Paris: Agence Francaise de Developpement, AFD.

Hilsum, Lindsey. 2015. "On the Border and in the Crossfire: Cameroon's War With Boko Haram." *The Guardian*. 13 May 2015. www.theguardian.com/world/2015/may/13/on-the-border-and-in-the-crossfire-cameroons-war-with-boko-haram

Hossain, Naomi, L. Brito, F. Jahan, A. Joshi, C. Nyamu-Musembi, B. Patnaik, M. Sambo, A. Shankland, P. Scott-Villiers, D. Sinha, D. Kalita, and N. Benequista. 2014. *Them Belly Full (But We Hungry) Food Rights Struggles in Bangladesh, India, Kenya and Mozambique*. Brighton, UK: Institute of Development Studies.

Hossain, Naomi, and Devangana Kalita. 2014. "Moral Economy in a Global Era: The Politics of Provisions During Contemporary Food Price Spikes." *Journal of Peasant Studies* (May):1–17.

Hossain, Naomi and Duncan Green. 2011. "Living on a Spike: How Is the 2011 Food Price Crisis Affecting Poor People?" www.oxfam.org/sites/www.oxfam.org/files/rr-living-on-a-spike-food-210611-en.pdf

IRIN. 2008. "Cameroon: Rapid Intervention Military Unit Strays From Its Mission." *IRIN News*. 29 August. www.irinnews.org/report/80065/cameroon-rapid-intervention-military-unit-strays-from-its-mission

Lappé, Frances Moore, Jennifer Clapp, Molly Anderson, Robin Broad, Ellen Messer, Thomas Pogge, and Timothy Wise. 2013. "How We Count Hunger Matters." *Ethics and International Affairs* 27 (3):251–259.

Legwegoh, Alexander F., Evan D. G. Fraser, Krishna K. C. Bahadur, and Philip Antwi-Agyei. 2015. "Do Dietary Changes Increase the Propensity of Food Riots? An Exploratory Study of Changing Consumption Patterns and the Inclination to Engage in Food-Related Protests." *Sustainability* 7:14112–14132.

Mason, M. N., T.S. Jayne, A. Chapoto, and C. Donovan. 2011. "Putting the 2007/2008 Global Food Crisis in Longer-Term Perspective: Trends in Staple Food Affordability in Urban Zambia and Kenya." *Food Policy* 36 (3):350-367.

Maxwell, Simon. 2009. "Africa: Global Crisis Will Hit Continent's Cities." *All Africa* 27 March 2009. http://allafrica.com/stories/200903270171.html

Minot, Nicholas. 2011. "Transmission of World Food Price Changes to Markets in Sub-Saharan Africa." www.ifpri.org/sites/default/files/publications/ifpridp01059.pdf

Moseley, William G., Judith Carney, and Laurence Becker. 2010. "Neoliberal Policy, Rural Livelihoods, and Urban Food Security in West Africa: A Comparative Study of The Gambia, Cote d'Ivoire, and Mali." *Proceedings of the National Academy of Sciences of the United States of America* 107 (13):5774–5779.

National Human Rights Observer. 2008. "Cameroon 25–29 February 2008: A Discreet and Bloody Crackdown." The National Human Rights Observer with Assistance from ACAT-Cameroon and ACAT France.

Observatoire National des Droits de L'homme. 2009. "Cameroun 25–29 fevrier 2008: Une répression sanglante a huis clos." Douala, Cameroon. http://tribunalarticle53.com/wp-content/uploads/2011/08/rapport_ondh_fevrier_2009.pdf

Observatoire National des Droits de L'homme. CCFD-Terre Solidaire. 2011. "Cameroun: Rapport sur la situation des droits de l'homme." http://ccfd-terresolidaire.org/IMG/pdf/rapportdroitshomme2008-10cameroun.pdf

Pinstrup-Andersen, Per. 2010. *The African Food System and Its Interaction With Human Health and Nutrition*. Edited by P. Pinstrup-Andersen. Ithaca: Cornell University Press and United Nations University.

Pritchard, David, and Francis Pakes. 2014. *Riot, Unrest and Protest on the Global Stage*. Basingstoke: Palgrave Macmillan.

Rao, P. 2014. "Food Crisis in Cameroon." *Africa Renewal* 28 (1):3. www.un.org/africarenewal/magazine/april-2014/africa-watch

Reuters. 2008a. "Biya Plan to Keep Power in Cameroon Clears Hurdle." *Reuters*. 8 April 2008. http://in.reuters.com/article/2008/04/08/cameroon-constitution-idINL0836856920080408

Reuters. 2008b. "Cameroon Ups State Wages, Cuts Prices After Riots." *Reuters*. March 8 2008. www.reuters.com/article/2008/03/08/idUSL08554176

Reuters. 2008c. "Cameroon's Biya promulgates law allowing third term." *Reuters*. 15 April 2008. http://uk.reuters.com/article/cameroon-constitution-idUKL1585217420080415

Sneyd, Adam. 2014. "Neopatrimonial African Capitalism? Conceptual Adventures via John Kenneth Galbraith." *Air and Space Power Journal – Africa and Francophonie* 5 (2):21–47.

Sneyd, Adam, Alexander Legwegoh, and Lauren Sneyd. 2015. "Food Politics: Perspectives on Food Security in Central Africa." *Journal of Contemporary African Studies* 31 (1):141–161.

Sneyd, Lauren Q. 2013. "Wild Food, Prices, Diets and Development: Sustainability and Food Security in Urban Cameroon." *Sustainability* 5 (11):4728–4759.

Sneyd, Lauren Q. 2015. "Zoning In: The Contributions of Buyam-Sellams to Constructing Cameroon's Wild Food Zone." *Geoforum* 59:73–86.

Sneyd, Lauren Q. 2016a. "A Missed Opportunity for Research and Development Interventions: Gender and the Forest Food Trade in Urban Cameroon." In *Transforming Gender and Food Security in the Global South*, edited by Jemimah Njuki, John R. Parkins, and Amy Kaler. Milton Park, UK: Routledge.

Sneyd, Lauren Q. 2016b. "Wild Food Consumption and Urban Food Security." In *Rapid Urbanization, Urban Food Deserts and Food Security in Africa*, edited by Jonathan Crush and Jane Battersby. Springer: Switzerland.

Sneyd, Lauren Q., Alexander Legwegoh, and Evan D. G. Fraser. 2013. "Food Riots: Media Perspectives on the Causes of Food Protest in Africa." *Food Security* 5 (4):485–497.

Steyn, Nelia P., and Zandile J. McHiza. 2014. "Obesity and the Nutrition Transition in Sub-Saharan Africa." *Annals of the New York Academy of Sciences* 1311 (1):88–101.

6 Demanding accountability for hunger in India

Anuradha Joshi, Biraj Patnaik and Dipa Sinha

Introduction

The new politics emerging out of the food, fuel and financial crises that started in 2008 was born out of a sense of discontent with the ruling classes. The roots of the protests in country after country, and the responses they evoked can be best viewed, as this book argues, through the lens of a 'politics of provisions,' – the way in which "common people interacted with their rulers over subsistence," and how these interactions were "permitted and shaped by pre-existing social and political networks, both among rioters and between them and their rulers" (Bohstedt 2016: 1036). Riots (or not) and responses (or repression) at moments of crises throw the ongoing politics of provisions into special focus by highlighting the boundaries of what the rulers will accept as a form of protest by the ruled, and what the ruled deem as necessary action.

In each of the country cases in this volume we have seen how this politics of provisions played out with particular nuances. The price spikes of 2008 and 2011 were the source of social unrest, with commodity-dependent democracies such as Cameroon, Mozambique and Kenya being particularly vulnerable to protests. India is far less dependent on food imports, and although it did experience rising inflation, it did not witness sudden inflationary spikes during this period (Chandrasekhar and Ghosh 2013). Protests around food were few and highly localized, mainly because India has had a large program of subsidized food for the poorest households through the Public Distribution System (PDS) since the 1950s. The only reported 'food riots' were in fact around perceived corruption in the PDS in West Bengal. Instead, between 2001–2014, India saw a rise in mobilization around a 'right to food' which was aimed at ensuring food security for all, especially the most marginalized, through guaranteed entitlements of basic food grains at subsidized prices. Significantly, such mobilization varied spatially across states. While a few states saw riots, in most states strong grassroots movements emerged and played a key role in the national right to food campaign.

In this chapter we explore the politics of provisions in the case of India.[1] We ask: a) why does popular mobilization take the form of riots in some locations and rights-based movements in others? b) how does popular mobilization draw upon the imaginary of the moral economy of hunger? c) what features of the environment matter, especially the role of the media in framing and shaping discourse enable mobilization? d) through what pathways do responses to mobilization lead to institutionalized accountability for hunger?

We explore these issues by tracing popular mobilization and subsequent responses around food insecurity first in two states – West Bengal and Madhya Pradesh – and then exploring their linkages to the national right to food campaign. West Bengal witnessed widespread popular unrest in 2007, in what were called 'ration riots,' in which rural communities directed public anger at the malfunctioning of the subsidized food distribution system in the state. The riots were violent, spontaneous and characterized by burning and looting of PDS shops by angry villagers. Yet, these places had never been locations of widespread acute hunger or malnutrition. By contrast, in Madhya Pradesh, malnutrition and hunger had become a big public issue in the mid-2000s, when the National Family Health Survey showed that it had the highest rate of malnutrition in the country (60 per cent of the children were malnourished at the time). Chronic hunger was a feature of life, especially in tribal areas. Rather than leading to rioting however, popular mobilization led by local grassroots organizations took the form of a rights-based movement against malnourishment and hunger, which linked up with the national right to food campaign.

How do we explain these contrasting trajectories of two states within the same national system? Our research finds that both riots as well as rights-based movements were framed in terms of an implicit breach of the social contract; a rupture in what Hossain and Kalita elaborate as the contemporary moral economy (Hossain and Kalita 2014). However, the causes of the protests, who participated in them and the form they took were different in the two states. Moreover, while both elicited responses from the state, only the rights-based movements had the potential to translate into durable responses to hunger. However, protest alone was not enough – the potential for gains at scale can be realized only if the political context is favorable and protestors can seize opportunities that arise. For rights-based movements, key enabling conditions included: strong networks that linked national policy actors with grassroots groups, thus increasing both legitimacy and strength at the national and local levels; active strategic use of the media; close and regular interactions between national and sub-national structures of government; and a framing of demands in terms of socially and widely accepted rights.

Methodology

The research reported here draws on two sources of evidence. First, at the national level, we constructed an event catalogue using media reports, and interviewed government officials, politicians and members of the national right to food campaign. The empirical material generated through these methods was supplemented by data from official reports, newspaper articles and other documents.[2]

Second, in each of the two states (Madhya Pradesh and West Bengal) we carried out focus group interviews with a purposive sample of villagers in two villages, supplementing those findings with semi-structured interviews with activists, journalists and relevant government officials at the state level.[3] We selected Bankura district in West Bengal, as it was one of the places where the 'ration riots' of 2007 had started, and we had prior contacts with members of the agricultural workers' union in that district.[4] We carried out focus group discussions and local leader interviews in two villages in Bankura: Radhamohanpur (the site of the first riots) and Barokumira (where riots erupted later). Within Madhya Pradesh, we chose Satna district because it is one of the districts with the highest level of malnutrition, has a large tribal population and has witnessed local grassroots action and mobilization on these issues. Moreover, through the state-level right to food campaign, we had direct access to the work of one of the grassroots organisations working with tribals in the chosen district – the Adivasi Adhikar Manch (AAM). Again, two villages were chosen for the focus group discussions: Chitehara (a tribal-dominated village) and Baraha Mawan (a mixed village with tribal and non-tribal populations).

A word about our positionality: of our research team, two members were prominent members of the right to food campaign. While on the one hand, this meant that it was easier to gain access to senior policymakers and activists at short notice; it meant, on the other hand, that interviewees often presumed that as part of the campaign, interviewers had particular positions on the issues under question and occasionally responded in ways that addressed the campaign rather than the interview questions. Despite our best efforts to mitigate against this through triangulation, making non-activists lead interviewers and starting interviews with a more critical view of the right to food movement, some bias inherent in interview context remains. The findings need to be read keeping these limitations in mind.

Food security in India

Between 2006 and 2013, India witnessed increasing food inflation but in contrast with global trends, there was no food price volatility. From 2008 onwards, food prices in India showed a rising trend compared to the two distinctive spikes in global food prices in 2008 and 2011. Further, in the Indian context, food prices were driven up not so much by prices of food grains (except for a brief period) but by other food articles such as pulses, vegetables and fruits (Chandrasekhar and Ghosh 2013). Considering the poor diversity in diets of the poor in rural India, food price inflation does not seem to have had much of an effect on their access to food. This was also a period that saw an overall increase in real wages (Drèze and Sen 2013, Gulati *et al.* 2014).

However, hunger and malnutrition were (and still are) widespread in India. Despite high economic growth, the decades of the 1990s and 2000s did not show an equivalent improvement in food consumption and malnutrition figures (Deaton and Dreze 2009; Svedberg 2008). The average calorie consumption decreased, with the poor continuing to consume much less than those in higher-income deciles. While there was some increase in consumption of other food

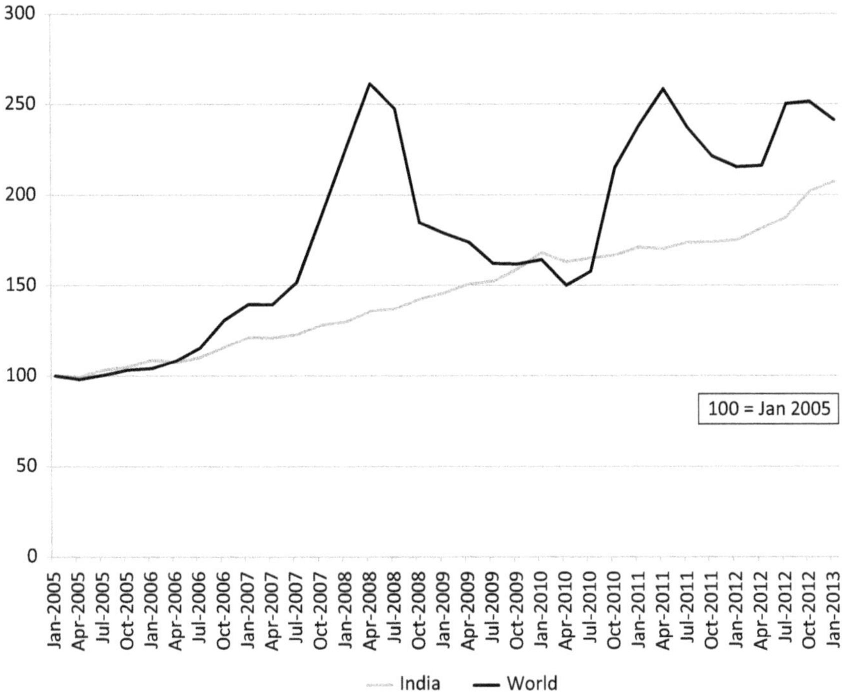

Figure 6.1 World and India cereal price indices, 2005–2013.

Source: Calculated from price index data available from FAO (for world) and Ministry of Commerce and Industry, Government of India (for India)

www.fao.org/worldfoodsituation/foodpricesindex/en/ accessed on March 25, 2015

http://eaindustry.nic.in/download_data_0405.asp accessed on March 25, 2017

items such as pulses, fruits and vegetables, they were still at extremely low levels. Available data show a slow improvement in nutrition status between the early 2000s and the early 2010s. The National Family Health Survey (NFHS-3) of 2005–2006 showed that 44 per cent of children under five were underweight and more than half of women of reproductive age were anaemic (Indian Institute of Population Studies 2007). The NFHS-4 for 2015–2016 shows some improvement in malnutrition, with 35.7 per cent of children under five being underweight (Indian Institute of Population Studies 2017). Improvement in nutrition levels in this period could be because of the efforts made by the central and state governments towards providing social security and enhancing food security through programs such as the employment guarantee scheme and the public distribution system (von Grebmer *et al.* 2014).

Beginning in 2004, a number of rights-based legislations and schemes were introduced. Among the most relevant to food and nutrition security were the

Mahatma Gandhi National Rural Employment Guarantee Act (MGNREGA) in 2005 and the National Food Security Act in 2013. The MGNREGA guarantees 100 days of employment at minimum wage to every rural household that demands work. While there have been implementation gaps, the Act has generated millions of days of employment and had a positive impact on rural wages, food security and migration (Afridi and Iverson 2014). The impact of the National Food Security Act was still being assessed at the time of this study. However, most of the entitlements under the Act had already been in existence since 2001 through orders of the Supreme Court under the 'right to food' case. Through the food schemes covered under the food security Act and the Supreme Court orders, a range of entitlements help people meet their food and nutrition security needs.

The most important of these is the Public Distribution System (PDS). The PDS was started by the British as a wartime food grain distribution system, and later revised into a welfare measure in the 1950s. Initially the program was universal; however, over the years it underwent a series of changes and by the 2000s, the government targeted food subsidies to the poorer sections of the population through the distribution of three levels of entitlement cards (in order of decreasing vulnerability): the Antodya Anna Yojana (AAY), the Below Poverty Line (BPL) cards and the Above Poverty Line (APL) cards (Swaminathan 2000). Under the scheme, households receive subsidized food grains (mainly wheat and rice), sugar and kerosene oil.

The aims of the PDS are twofold: to maintain stability of prices through government purchase of food grains at a minimum support price for farmers, and to distribute the food grains purchased at a subsidized rate to cardholders. The distribution occurs through a network of licensed Fair Price Shops, who hold a list of all the cardholders who are entitled to purchase the subsidized grain from their shops. Cardholders in turn present their cards to purchase their entitlements. The central and state governments are both involved in the PDS. A formula is used to assess how much subsidized food grain each state is entitled to receive, and states have to request the central government to send the amounts they need up to the entitlement limit, and the central government then distributes the grain through its stores. Thus, the actual availability of food grains in shops depends upon a certain level of central and state government interactions and a complex distribution system.

Other schemes addressing food security include the school midday meal scheme providing cooked meals to children in government schools, and the Integrated Child Development Services scheme providing supplementary nutrition, growth monitoring, nutrition counselling and pre-school education of children under six. Supreme Court orders in 2001, 2004 and 2006 directed that these schemes be made universal entitlements. These orders saw a tremendous expansion in these schemes, which currently cover about 120 million school-going children and 70 million children under six years of age.

The quality of implementation of these schemes varies greatly across states (Kabra and Ittyerah 1992, Jishnu 2004). Criticisms of the PDS are particularly trenchant and relevant here. The PDS is considered highly corrupt, and shop owners are believed to divert subsidized grain to the open market to pocket the subsidies, while reporting the grains were not received, or not in the required quantity. The allocation of cards is also contentious: eligible households may be excluded while some ineligible households are included. Moreover, cardholders find it difficult to buy their entitlements in instalments even if this is formally allowed (Programme Evaluation Organisation 2005; Khera 2011).

Most food-related government programs are centrally funded (sometimes with matching state funds), with the implementation left to state governments. For example, all procurement and storage of food grain is primarily under the control of the central government through the Food Corporation of India (FCI). While the central government allocates the grain to the states, it is the responsibility of the state to ensure it reaches beneficiaries. Further, state governments have few sources of revenue of their own, and many have faced serious fiscal crises in the last two decades. Therefore, while there can be wide differences in the way schemes are delivered, there is little room for expansion or creation of new schemes at the state level. Even then, some state governments have provided social support beyond that allocated by the central government (Patnaik and Sinha 2016).

The politics of popular mobilization around social provisioning

New social movements emerging in India in the 1970s focused on issues such as environment, gender and housing to fill a vacuum left by political parties and older movements, and as a critique of the state and its inability to deliver basic entitlements for women, lower caste groups and the poor (Kothari 1984, Omvedt 1994). The neoliberal policies followed by the Indian government since the 1990s met with firm grassroots resistance (Baviskar and Sundar 2008). Civil society organizations that emerged out of these new struggles played a key role in expanding the democratic obligations of the state and creating expectations among the poor. Thus, the decades of the 1990s and 2000s saw the emergence of a number of networks, campaigns and social movements built around socio-economic rights, e.g., the Right to Education Campaign, the People's Act for Employment Guarantee, the Right to Health movement (Jan Swasthya Abhiyan), the Campaign for Survival and Dignity (which demanded the Forest Rights Act), the National Campaign for People's Right to Information and the Right to Food Campaign. One of the distinctive features of the campaigns was the simultaneous deployment of protest and collaboration strategies at different levels (Pande 2014).

In this context, 2004 was a watershed: the ruling coalition, the National Democratic Alliance led by the Bharatiya Janata Party, was overthrown in a landslide victory by the Congress party-led United Progressive Alliance. The ruling coalition had gone to the electorate with images of a prosperous urban India that included wide roads and shopping malls using "India Shining" as its main campaign slogan. By contrast, the principal opposition party – the Indian National Congress – claimed to represent the *aam aadmi* (common man), highlighting the neglect of rural India, unemployment and the need for social security. Even though the broad economic policies of the two parties were not very different (in fact, it was a Congress government that initiated neo-liberal economic reforms in 1991), the surprise victory of the Congress-led alliance was widely interpreted as a mandate for explicitly pro-poor policies.

The alliance was able to form a government with a clear majority with the support of the Left Front (consisting of the communist parties), who between them had about a tenth of the parliament seats. The Left Front offered support on the basis of a National Common Minimum Programme which included many of the issues taken up by the rights-based campaigns such as the employment guarantee legislation, strengthening of the PDS, greater investments in education and health and so on. The program agenda was strengthened by the creation of a new body called the National Advisory Council headed by the president of the Congress party and including a number of activists who were part of rights-based campaigns and movements. This inside-outside position of the rights-based campaigns – demanding rights through protests outside the state while simultaneously pushing the rights agenda from within the Advisory Council, is one of the hallmark features of India's politics of provisions in the period from 2004–2014. As we will see later, linking protesters in the village to the corridors of policymaking in the capital through various points of access is one of the reasons for their relative success in institutionalizing entitlements (Khera 2013).

Riots in West Bengal and rights in Madhya Pradesh

In 2007, in a village in Birbhum district of West Bengal, APL cardholders angry about being denied subsidized wheat due to them under the PDS forced the local shop-owner to sell wheat at a lower price. The spontaneous mobilization soon spread and there were riots around PDS shops in several villages in the region. One shop-owner committed suicide; the police fired on crowds injuring several people, and shops were burned and looted (Bhattacharya 2008; Majumdar 2007). In West Bengal, the ruling left party coalition, led by the Communist Party (Marxist) (CPM), had been in power for over three decades. With party units in every village; local government, teachers' unions, and PDS ration-dealers' associations all politicized; spontaneous riots targeting PDS shops were an unusual occurrence.

The riots were neither initiated by the poorest villagers, nor about hunger or malnutrition; rather they were about accessing wheat (not the staple food in West Bengal) under the PDS, by Above the Poverty Line (APL) card-holding households. The proximate reason for the riots was the difference between the price of open market and the price of the subsidized wheat. For a decade previously, the APL prices of wheat had only been slightly lower than market prices, leading to lower uptake. In 2006 central government had decided to limit allocations under the APL category to the average taken up by states in the previous three years. At the same moment, however, the price of wheat began to rise, making the PDS wheat relatively more valuable. Given that the shop-owners and their association were closely linked to the ruling party, the state was seen as siding with corrupt shop-owners and protecting them.

This was also a period when the hegemony of the Left Front was beginning to weaken. Massive protests had broken out across the state in response to the state government's decision to acquire hundreds of acres of agricultural land for a private car factory, Tata Motors. The main opposition party in the state – the Trinamool Congress – seized the opportunity offered by the widespread discontent and assumed leadership of the protest movement. The state government responded with police action, but was eventually forced to withdraw the acquisition plans and return the land to the farmers (Banerjee 2011; Banerjee and Roy 2007). These larger political developments provided the 'environmental mechanisms', against which the PDS rioters leveraged general discontent with the Left Front government (McAdam, Tarrow and Tilly 2001).

The popular view was that PDS shop-owners were generally corrupt and that both inclusion and exclusion errors were pervasive. As one villager noted: "APL-BPL divide is not correct. The poorest people who should be BPL don't have BPL cards. There is a political reason." Another said: "some dealers are dacoits [armed robbers] and some are thieves."[5] So although the immediate issue concerned the unavailability of APL wheat, both APL and BPL families participated in the riots. In Barokumira village, villagers dragged the shop-owner to a village meeting and made him accept that he had been charging cardholders above the official price. To compensate for his wrongdoing, he was asked to contribute 200,000 rupees toward village development. Unable to pay, three days later he committed suicide. Eight villagers had cases filed against them, but in the end, they were released.

Deep discontent with the way in which the ruling party had an iron grip on local governance also underpinned the riots. In Radhamohanpur village on September 16, 2007, the CPM was holding a zonal convention. Rumor had it that the CPM was part of the corruption in the PDS. Villagers decided to confront the shop-owner in front of other party leaders at the convention, to create pressure for reforms. However, when villagers went to the zonal meeting, the CPM protected the shop-owner. Violence broke out and police were brought in to control the protests. Ultimately several people were arrested and a curfew imposed on the village. Villagers expressed their discontent with political parties, which in West Bengal at the time, was synonymous with state government. As one landless woman observed:

> When the (political) parties came to ask for vote, I said "what is that we get to vote for you?" I gave them a warning, "do not come for vote to us . . . what is that you give me? What is it that you help us with?" We do not even eat or wear their given stuff). We work and that is why we can eat a fistful . . . In every household, they will get seven to eight votes, that is how they see us. Why shouldn't we speak up?[6]

In these incidents, villagers framed their demands as a '*janagana andolan*' (people's movement), yet recognized that the mobilization had limited results. As one villager put it, "we went to ask for wheat, and 'ate' bullets." The political party in collaboration with the shop-owners association agreed to compensate the two people who were shot, but little else changed. However, they felt the *khadyo andolan* (food movement) would continue if necessary – "we have no choice, but yes. If we don't have kerosene oil, how will we light our houses? We don't even have money to buy candles. There will be a struggle again."[7] In West Bengal, then, the focus was on rights, but narrowly, in the sense of the right to food through entitlement programs.

In Madhya Pradesh, the release of the National Family Health Survey (NHFS) data in December 2006 highlighted that the state not only had the highest level of child malnutrition in the country, it also saw an increase in the prevalence of malnutrition since the previous survey (1997–1998) (Indian Institute of Population Studies 2007). At the same time, a number of deaths due to malnutrition were reported in the media, bringing the issue of hunger and malnutrition to national attention. Simultaneously, grassroots groups like the Aadivasi Adhikar Manch (AAM), were mobilizing villagers to demand better child care centers, improved delivery in the Public Distribution System and the establishment of nutrition rehabilitation centers. In collaboration with the state and national level right to food campaign, AAM was able to convert popular mobilization into sustained pressure for improving entitlements for the poor.

The story of popular mobilization and its effects in Madhya Pradesh has to be understood in terms of the ways in which local groups have managed to leverage activism at higher levels as well as creating allies within the media to gain traction on the broader issue of sustainable livelihoods. These provide the environmental mechanisms with which contention can unfold (McAdam, Tarrow and Tilly 2001). The AAM, set up in the late 1990s by two activists who had been trained by a well-established non-governmental organization (NGO), initially started working on land rights and were successful in retaining land in de facto control of the tribal population. In Satna district, poverty and food insecurity are widespread, and the poor depend upon subsistence farming and forest produce. One villager described the food security situation as "four months of surplus and the rest a period of want."[8] Government programs such as PDS and midday school meals are critical for survival.

AAM's links with a media advocacy organization working on the issue of child malnutrition – Vikas Samvad – led to AAM focussing on malnutrition and greater visibility for the issue. Between 2007 and 2012, AAM organized protest marches and public hearings at local and district levels, provided evidence of malnutrition from the grassroots, and participated in state and national protests led by the right to food campaign. In November 2009, 30 activists from AAM participated in a national rally in Delhi to reject the minimalist draft of the Food Security Act proposed by the national government. AAM connects child malnutrition to broader grievances which resonate with the population, such as malfunctioning government programs, increasing unviability of agriculture due to high input prices and inflation, and the lack of employment opportunities.

AAM also mobilized locally on policy issues in line with the national position of the right to food campaign. These local demands included universalization of the PDS, inclusion of pulses and edible oils within the PDS and improving the quality of grain provided. One of the public hearings on malnutrition was attended by a member of the National Commission for the Protection of Child Rights which is affiliated to the right to food campaign. The relative gains made by AAM locally were partly a result of these links to higher levels of activism and the media. After one of the protest marches, one tribal activist recalled:

> After that all kinds of officials who we had never seen before descended on our village. Those who did not have the ration cards were issued the cards. All of us were given food grains and everything. There was a medical camp in the village for fifteen days. All the sick children were taken care of and things became better.[9]

The presence of national-level influential people who witnessed first-hand the testimony of the poor about child malnutrition led to the opening of a nutrition rehabilitation center. Since 2010, mobilizations have declined, partly because most of the demands for direct interventions on malnutrition have been met.

The cognitive frame of those who participated in the struggles was one of rights. Describing the protests, a member noted, "It was almost like an election atmosphere where the contradictions to our interests got clearly highlighted and people were ready to oppose those who obstructed their rights."[10] Protest slogans reflected this rights-based perspective: 'government land is our land.' When asked if they were worried about state repression over their participation in the protests, one of the participants replied, "How will they arrest us? It was a rally." The mobilization by AAM activists has increased awareness of rights, emboldening poor villagers vis-à-vis their state. As one villager put it, "Now whenever anyone comes asking about our problems, we know how to present our case to the government and any other journalists." The rights that villagers refer to here are broader than accessing food entitlements.

The responses to the popular mobilization in West Bengal and Madhya Pradesh were very different and owe a lot to the political moment within which protests took place. In Madhya Pradesh, the response was twofold: the creation of a high-level structure to converge action around malnutrition in the form of the inter-departmental Atal Bal Mission for child health and nutrition, set up under the leadership of the chief minister; and increase in the number of PDS beneficiaries.[11] The Mission was comprehensive, looking at production and distribution of food grains as well as the social determinants of malnutrition, including water and sanitation, decentralization, and coordination of Women and Child Development, Health and Family Welfare, and Rural Development and Panchayati Raj ministries.

In West Bengal, the policy response around a year later, and was primarily the provision of a state subsidy to reduce the prices of rice and wheat available to families living below the official poverty line. But this was less an immediate response to the riots as it was to the elections, which were less than 100 days away when the scheme was announced. A senior Food Ministry bureaucrat pinned the blame for the crisis in West Bengal on local factors and corruption in the state machinery, and not on the reduction of the APL quota for the state, attributing the crisis to large-scale hoarding by private actors within the PDS network:

> In almost all other states, there is a state government entity, which picks up food grains from FCI godowns and that entity keeps the food grains in their own godowns and then distributes it to fair price shops. In both Kerala and West Bengal the job of a wholesaler is done by private people. They are called authorised wholesale dealers. Ironically, in both the left-ruled states it is the private dealers who lift the grains from FCI. So we have tried and failed to persuade the states to set up their own corporations and do this work.[12]

The policy response in West Bengal thus focused on a populist price intervention that was easier to undertake than more difficult structural reforms in the management of the PDS.

The case of Madhya Pradesh is more representative of what was happening elsewhere in India, particularly in the links between localized efforts and national policy advocacy by the right to food campaign. Supreme Court orders were disseminated by the state-level campaign to the AAM, which used them in their mobilization strategies. The networked nature of the AAM gave it power well above its modest organisational base. The local media, linked to and feeding the national media, were heavily involved in highlighting malnutrition, turning it into a political issue, as evidenced in state assembly debates. In Madhya Pradesh, while individual PDS shops might have been linked to particular political patrons, there was no state-wide capture of the PDS system by any political party, so mobilization for food security and malnutrition was not viewed as being against any particular political party. The AAM campaigns addressed all political parties in an effort to highlight the demands of the people, the main ones being access to and expansion of services related to food and malnutrition. The mobilizational repertoires were common ones: of *dharnas* (peaceful demonstrations), petitions, *jan sunwais* (social audits) and rallies.

The difference in policy response in each case was driven by the underlying nature of the problem but also by its popular articulation. In Bengal, the riots were essentially 'ration' riots linked directly to the PDS, and the response was restricted to policy changes in PDS prices and beneficiary numbers. In Madhya Pradesh, the popular mobilization was around broad food security issues and specifically malnutrition, and the policy response was larger and more institutional, with influence on the national policy debates.

The Supreme Court and the right to food case

Inflation, and especially food price inflation, has been an issue of national public debate since the early 2000s, and by 2008 was a pressing concern. The initial response of the government to inflation was narrow, revolving around monetary policy in respect of banning rice exports and futures trading in four commodities presumed to affect domestic prices (De Schutter 2010).[13] A more direct government response to issues of hunger and access to food was forced by three interlinked factors: an activist Supreme Court, the rising political salience of food security in the media and continuous pressure by the right to food campaign.

The Supreme Court played an active role in shaping government responsibility for hunger through decisions on the People's Union for Civil Liberties (PUCL) vs. Union of India and others, Civil Writ Petition 196/2001 (popularly known as the right to food case) (Banik 2016). The case revolves around a petition filed by the PUCL, a leading Indian human rights organization, in the Supreme Court. This argued that in the context of successive droughts in 2001, it was imperative that the government use the accumulated food stocks of approximately 50 million tons to mitigate drought relief and hunger. The petition went beyond situations of drought and focussed on the need to uphold the 'right to food.' In ruling on the case, the Supreme Court laid out path breaking jurisprudence (De Schutter 2013).[14] The orders converted all food and employment schemes run by the central government into legal entitlements, universalising food entitlement programs for children, and instituting an independent accountability mechanism, the Commissioners to the Supreme Court, to monitor all food and employment programs.

Supreme Court activism ensured that right to food issues became mainstream in the public discourse. Media coverage forced the government to act under pressure of judicial orders and the public debate that ensued.[15] An illustration of this is the reversal of state policy linking the official poverty line to targeting of food security schemes. During 2011–2012, the Supreme Court made a series of observations on the low threshold of the poverty line (24 rupees per capita per day for rural areas and 32 rupees per capita per day at 2010 prices) set by the Planning Commission for targeting in food schemes, including the PDS. The low threshold of the poverty line was picked up by the print, electronic and social media and debated extensively.[16] The barrage of reporting finally prompted the Minister of Rural Development and the deputy chairperson of the Planning Commission to issue a press statement de-linking the poverty line from the targeting of anti-poverty schemes (Government of India 2012).

By 2009, then, the right to food had been firmly mainstreamed in national discourse, and both major political parties included it in their election manifestos. The passage of the National Food Security Act in 2013 was evidence of how far the discourse had moved toward inclusiveness. A comparison of the manifestos of the two major political parties both for the Assembly elections in 2008 and the national elections in 2009 shows that many of the demands met through the Act had been election promises. Both the major parties promised subsidized food grains and an expanded PDS. Notably, both manifestos mention a universal PDS and community kitchens for the urban poor. The Indian National Congress manifesto commits to the universalization of the Integrated Child Development Scheme and for the first time mentions a 'Right to Food Act.'

This focus on the PDS in the national elections came against the background of an improvement in the PDS in many states in the period preceding the 2009 elections and within the context of high food prices. Chhattisgarh had state assembly elections in 2008, and pioneering reforms in the PDS took place including a host of measures to increase transparency and accountability and reduce corruption. It was widely acknowledged that this contributed to the popularity of the Chief Minister and his party. Chhattisgarh became the first state outside the southern states to show that it was possible to improve the PDS and reap political gains from doing so. This period saw a host of other states also taking similar measures to increase PDS coverage and subsidies using their own revenues. It was therefore no surprise when the PDS became a major campaign pledge for both major parties.

The right to food campaign, which emerged out of the Supreme Court case, was also one of the principal agents of policy change in the period, including campaigning for the National Food Security Act. Between 2009 and 2013, the bill went through various drafts and amendments. The campaign prepared an alternative draft[17] and at each stage mobilized public opinion and engaged with the press, politicians and policymakers to expand the scope of the bill. A study of the various changes in the drafts reflects the richness of the policy discourse during the reference period (Sinha *et al.* 2014). For instance, the March 2010 draft of the Act by the Empowered Group of Ministers restricted benefits to 25 kg of rice/wheat a month at three rupees per kilogram for a section of the Below Poverty Line population (as defined by the Planning Commission). The second draft coincided with public debates about the poverty line, high inflation and lobbying by the right to food campaign, and entitlements were expanded to 35 kg for the poorest households, and coverage of 67 per cent of all households with 5 kg per person per month, with highly subsidized food grains. Other entitlements related to children and women also found a place in the final food security legislation, albeit in a diluted form (Sinha *et al.* 2014).

The final legislation, passed in 2013, guarantees subsidized grains to 820 million Indians. It provides for free midday meals to 130 million school children and supplementary nutrition to all children in the country below the age of six and guarantees a maternity entitlement to every pregnant and lactating mother. With an annual budget of nearly USD $25 billion, the National Food Security Act is one of the largest food programs in the world. As we saw, its final shape was the result of a long trajectory of mobilizing, contestation and debate between the campaign, academic observers, politicians and bureaucrats.

It would be remiss to overlook the contextual features that have enabled the gains made by the right to food campaign. First, the liberalization of the Indian economy in the early 1990s has consistently triggered civil society protests against the elevation of private above public interests (for example, protests against land appropriation for Special Economic Zones). Rising food prices have been viewed as part of the broader impacts of liberalization. Second, the precedent of state obligations to protect the poorest from food insecurity had been well-established since the early 1960s. Third, the 2004 elections in which the Congress-led UPA unexpectedly won opened the door to the possibilities for several social protection policies. The National Advisory Council formed soon after the election with

Sonia Gandhi as chair had political clout, and with a membership consisting of several representatives of social movements, was able to support several pieces of progressive legislation – including the MGNREGA, the Forest Rights Act and the Right to Information Act. The impact of these was critical to support for the Congress-led UPA in the 2009 elections, in which the PDS became a political issue. Several state governments reduced PDS prices and the NSFA appeared in the election manifesto of many parties. Moreover, by continuously passing orders compelling the government to expand and improve food-related programs such as the midday meal in the right to food case, many in government believed that the Supreme Court was overstepping its mandate and interfering in governance. Returning debates about entitlements to food to the parliament through the Right to Food Act was seen as placing policy and implementation back to its rightful place: the legislative sphere.

The politics of provisions

Were these politics of provisions rooted in an implicit social contract around protection against hunger? The evidence supports such an interpretation. Contention, by and large, was not in the form of riots but in the form of an expanded repertoire of engagement with the state through protests, social audits and media campaigns. Underpinning these was an expectation of justice, fairness and solidarity on the part of protestors and within government. The prevailing norm in India has always been that it is the state's responsibility to provide for the poor and protect them from the market, as exemplified by the mammoth food market intervention that is the PDS. Yet the roots of these expectations, the form they take and the responses they elicit, vary across the country. Thus, protests in Madhya Pradesh, or riots in West Bengal, and the broader right to food campaign must be seen in light of the expectations people have of their relationships with the landed elite, the community more broadly and the longer cycles of relationships between the state and its citizens.

Two points are of particular importance. The first relates to the erosion of patronage ties with landlords and other community elites and the rural poor. In many isolated and underdeveloped areas of India, the state has not been a significant presence and the landed elite have been the source of social protection. However, these relationships are rapidly disappearing. With declining access to traditional products from the forests, the de-linking of employment and social security due to migration, the changing nature of employment and more recently the advent of the NREGA, the roots of popular mobilization can no longer be tied to the moral economy of elites and landlords.[18] While exploitative landlords have not disappeared, their hold has weakened in many parts of the country. Seemingly in contradiction, it appears easier to direct anger against a 'faceless' anonymous state than an identifiable local personage who might subject one to reprisal. Patronage ties have in general been replaced by electoral accountabilities.

The second is that rights struggles in rural India have to be seen within a long-standing politics of provisions in which the Constitution has within it accepted a 'right to life' which has been drawn upon to make arguments in a variety of progressive cases, including the Supreme Court Right to Food case. Past histories of famine (including the Bengal famine), and the fact that hunger has always been an emotional and political issue, suggests that mobilization against hunger and malnutrition, no matter what form it takes, is more naturally directed against the state. The state on its part has been extremely sensitive to the threat of hunger and famine, as, if anything, its legitimacy rests of preventing starvation.[19] The colonial roots of the Bengal famine of 1943–1944, and its evocation in the struggle for independence give widespread hunger (or the risk of it) particular political salience. This is reinforced by the fact that civil society mobilization, under the auspices of the right to food campaign has managed to create an environment where the state is widely seen as responsible for addressing hunger.

These themes emerged from national level interviews and the state-level case studies. In Madhya Pradesh, the important rural social divisions are between the landed and the landless. Although awareness of state presence appeared to be relatively recent and partly an outcome of the mobilization work by AAM, there has been little expectation from patronage ties in the recent past (dependence on landlords ended at least two decades ago). Rather, prior to AAM's interventions, tribal populations had been left in a vacuum, squeezed between large scale economic changes that removed traditional sources of security and the absence of the – albeit limited – welfare of the state. The AAM was able to use 'moral economy' arguments of state responsibility to mobilize the poorest groups and demand accountability from the state.

In West Bengal, by contrast, social cleavages are more political than economic, in terms of those supporting the CPM and the rest. Shaped by the history of the left movement and the Left Front government that remained in period from 1977–2011, the moral economy has been always framed by expectations from the state. The PDS, which was the focus of the riots, had worked relatively well in providing for the poorest (BPL quota), as well as providing the allocated quotas of rice, the staple food. The problems in 2007, as we saw, were in the provision of wheat grains for those above the poverty line (APL) which led to the spontaneous protests.

In this context, where the state accepts responsibilities to address food insecurity, the role of the Public Distribution System (PDS) looms large. The PDS, by providing subsidized basic grains and fuel (even when functioning poorly), is viewed as the institution through which food security is to be ensured. This is so not only in the eyes of the poor, but also in the perceptions of the state, for whom intervening in food prices has been limited (e.g., open market sale of procured grain and anti-hoarding policies). The PDS has sown the seed for a growing acceptance of a 'right to food' whose duty bearer is the state.

During (and slightly after) the period 2006–2011, there was a lot of policy action around food prices and food security. However, policy response during this period has to be viewed as coinciding with the prevailing public debate on inflation and world food prices (and the related debate on the poverty line), the pressures exerted by the right to food campaign, the debates on the legislation of the National Food Security Act and the discussions in national media around the high levels of malnutrition in the country. As one right to food activist put it:

> We know that price is something that is not at all in the hands of, or control of poor people – that is why a PDS which could give extremely subsidized and adequate cereals, pulses and cooking oil . . . that is why our response was in a comprehensive way in terms of a campaign for a FSA – a kind of guarantee.[20]

In addition, 2009 was an election year during which issues of food had become politically salient. The whole period was characterized by the rise of a rights discourse, increasingly accepted by the government in power as part of realisable policy options.

The perceptions of policymakers are of relevance. Our interviews suggested that state officials do not view themselves as highly constrained by global markets, partly because they operate a large-scale procurement program, and partly because India's size and food production partially insulates it from world markets. A senior member of the Planning Commission noted, "in other countries the idea that world prices matter, that governments should do something about it is an issue, in India it is not."[21] The implication was that India was less affected by food prices, but also that there was never any doubt that the government was responsible for securing affordable food. However, policy responses driven by Supreme Court orders did not automatically translate into action at the grassroots level: they had to be activated by campaign groups on the ground to mobilize people to make accountability demands. What some policymakers seem to argue is that the right to food campaign's mobilizing was perceived as more acceptable to the bureaucracy because the right to food campaign offered 'solutions' and worked on details of programs.

This observation held in our two state-level cases. In West Bengal, although there was some immediate response to the riots, there were few institutional responses – the riots did not trigger state-wide improvements to the PDS, and key responses are more properly seen as preparation for the forthcoming elections that were anticipated to be hotly contested. Neither did the opposition party, which used the mobilisation for its own political ambitions, offer policy changes to the PDS or other related food security programs. By contrast, in Madhya Pradesh, the responses in the form of improvements in the Integrated Child Development Scheme, the creation of nutrition rehabilitation centers and the better functioning of the PDS were institution- and state-wide. The implication is that the progressive responses by the state in Madhya Pradesh are also more sustainable.

Conclusions

Bohstedt argues that 'politics of provisions' – "the ways in which common people interacted with their rulers over subsistence, and that those interactions were permitted and shaped by pre-existing social and political networks, both among rioters and between them and their rulers" – help explain why faced with a crisis of subsistence (food), people riot (or not) and states respond (or not) (Bohstedt 2016: 1036). In this chapter, the Indian case develops the argument further, showing that the politics of provisions are thrown into light not just in periods of crisis, but also in ongoing conditions of chronic food insecurity. Food riots are not simply a response to hunger, but symbolize feelings about an unjust economic and political system that coalesce around the issue of food.

The contrasts between the two states reinforce arguments about the importance of political economy to social and state reactions to food insecurity. Different political opportunities and different local institutions resulted in different forms of mobilisation. In West Bengal, the ruling Left Front had addressed issues of subsistence over three decades of socialist rule and resulted in a deeply politicked welfare system; in Madhya Pradesh activists, in collaboration with the right to food campaign, were starting to make inroads into problems of malnutrition in a context where the state was sensitive to such pressure. And yet the broader political and economic imaginaries were the same in the two states as well as at the national level for both activists as well as policymakers. Ideas that India was not a price-taker in food markets; that the poor were defenseless against sudden price rises; that elected governments were vulnerable to protests if food became unaffordable were common currency in networks and negotiations between the right to food campaign as well as those responsible for PDS and food policy. The very existence and scale of the PDS over six decades shows that social protection policy was well entrenched in the visions of all political parties. The language of protest, campaigns and riots was well understood in the long trajectory of social movements in India since Independence. Collectively, the Indian political economy of food combined with the political popular imagination of responsibilities combine to explain a period of ongoing chronic food insecurity and rising prices set within a broader politics of provisions, which led to the National Food Security Act of 2014. Whether this Act delivers on its entitlement promises for all will depend upon continuous mobilisation and responses set within the changing political and economic context.

Notes

1 The work presented here was funded by the ESRC project on the 'Moral and Political Economy of Accountability for Hunger' (ESRC – DFID Joint Research Programme). We are grateful to our interviewees for generously giving their time for this research, as well as Vaibhav Raaj and Shreya Bhattacharya who assisted in conducting the field work and the initial analysis. In addition, we are grateful to the RTF campaign for giving us access to campaign material. Prof. Jayati Ghosh, Dr. Santhosh Matthew, Kiran Bhatty, Dr. Santosh Mehrotra, Nitin Sethi, Anuradha Raman and Neelabh Mishra provided valuable comments at critical points in time. We are also grateful to

Anuradha Talwar, Sachin, Anand and Prateek for facilitating our site visits. We thank Naomi Hossain and Patta Scott-Villiers for useful comments on earlier versions of this chapter. The usual disclaimers apply.

2 The researchers had the privilege of access to the numerous grassroots organizations which constitute the Right to Food Campaign in India. Many of these organizations maintain a comprehensive database on food-related events in their area of activity. In addition, the team produced an event catalogue, mentioned in Chapter XX.

3 For details of each of the cases, see Sinha *et al.* 2014.

4 The union works in the district but was not part of the food riots, and did not have a direct presence in the villages that were most affected.

5 From a focus group discussion with senior citizens in Bankura District, August 29, 2013.

6 From a focus group discussion with agricultural laboring women in Bankura District, August 29, 2013.

7 From a focus group discussion with protestors in Bankura district, August 29, 2013.

8 From a focus group discussion with landless indigenous people in Bankura district, August 29, 2013.

9 Interview with agricultural laborer and activist in Madhya Pradesh, June 15, 2013.

10 From a focus group discussion with activists in Madhya Pradesh, June 15, 2013.

11 A comparison of the manifestos of the two major political parties in Madhya Pradesh – the ruling BJP and the Congress party – clearly shows that malnutrition and the right to food had emerged as dominant concerns.

12 Interview with author, June 2013.

13 Personal interview, senior member Planning Commission, New Delhi, September 18, 2013.

14 What is notable is that the *only* short-term intervention of the Indian government to liquidate food stocks through the PDS came about in 2011, only after repeated directives from the Supreme Court, to release at least 5 million metric tons of food grains from the stockholding to the PDS. The orders were made in the context of media reports of rotting food grains in the government storage facilities even as inflationary pressures on food continued unabated.

15 For a few examples see the following links: http://articles.timesofindia.indiatimes.com/2011-09-22/india/30188949_1_bpl-list-bpl-cards-affidavit; http://articles.timesofindia.indiatimes.com/2011-09-21/india/30183991_1_poverty-line-planning-commission-ample-proof; www.ndtv.com/article/india/supreme-court-slams-centre-planning-commission-over-criteria-for-bpl-population-100058; www.business-standard.com/article/opinion/is-the-bpl-census-correctly-structured-111081700029_1.html; www.dnaindia.com/india/report_rs25-a-day-is-enough-for-roti-kapda-shiksha-thinks-planning-commission_158965

16 For a few examples see the following links: http://articles.timesofindia.indiatimes.com/2011-09-22/india/30188949_1_bpl-list-bpl-cards-affidavit; http://articles.timesofindia.indiatimes.com/2011-09-21/india/30183991_1_poverty-line-planning-commission-ample-proof; www.ndtv.com/article/india/supreme-court-slams-centre-planning-commission-over-criteria-for-bpl-population-100058; www.business-standard.com/article/opinion/is-the-bpl-census-correctly-structured-111081700029_1.html; www.dnaindia.com/india/report_rs25-a-day-is-enough-for-roti-kapda-shiksha-thinks-planning-commission_158965

17 The alternative draft, and a number of petitions and letters from the campaign to the government in relation to the NFSA, are available on www.righttofoodindia.org

18 This point was made repeatedly in focus group discussions in Satna, Madhya Pradesh.

19 The recent heated debate about definitions of the poverty line illustrate the extent to which poverty (including food security) and related relief programs are high-stakes issues.

20 Interview with author, June 2013.

21 Personal interview, senior member Planning Commission, New Delhi, September 18, 2013.

Bibliography

Afridi, F., and V. Iverson. 2014. "Social Audits and MGNREGA Delivery: Lessons From Andhra Pradesh." *India Policy Forum*, National Council of Applied Economic Research, 10 (1):297–341.

Banerjee, P. S. 2011. "Party, Power and Political Violence in West Bengal." *Economic and Political Weekly* 46 (6):4–11.

Banerjee, P. S. and D. Roy. 2007. "Behind the Present Political Unrest in West Bengal." *Economic and Political Weekly* 42 (22):2–8.

Banik, D. 2016. "The Hungry Nation: Food Policy and Food Politics in India." *Food Ethics* 1 (1):29–45.

Baviskar, A., and N. Sundar. 2008. "Democracy Versus Economic Transformation?" *Economic and Political Weekly* 43 (46):87–89.

Bhattacharya, D. 2008. "The Politics of PDS Anger in West Bengal." *Economic and Political Weekly* 43 (5) 63–69.

Bohstedt, J. 2016. "Food Riots and the Politics of Provisions From Early Modern Europe and China to the Food Crisis of 2008." *Journal of Peasant Studies* 43 (5):1035–1067.

Chandrasekhar, C. P., and J. Ghosh. 2013. "The Changing Pattern of Food Inflation in India." *Macroscan*. www.macroscan.net/index.php?&view=article&aid=1571

De Schutter, O. 2010. "Food Commodities Speculation and Food Price Crises: Regulation to Reduce the Risks of Price Volatility." Briefing Note 2. http://www2.ohchr.org/english/issues/food/docs/Briefling_Note_02_September_2010_EN.pdf

De Schutter, O. 2013. "Interim Report of the Special Rapporteur on the Right to Food." 68th Session, United Nations General Assembly. www.srfood.org/images/stories/pdf/officialreports/20131025_rtf_en.pdf

Deaton, A., and J. Dreze. 2009. "Food and Nutrition in India: Facts and Interpretation." *Economic and Political Weekly* 44 (7) 42–65.

Drèze, J., and A. Sen. 2013. *An Uncertain Glory: India and Its Contradictions*. Princeton, NJ: Princeton University Press.

Government of India. 2012. Joint statement by Montek Singh Ahluwalia, Deputy Chairman of the Planning Commission and Rural Development Minister Jairam Ramesh, 4 October 2011. Accessed August 2013. www.ndtv.com/article/india/rs-32-controversy-joint-statement-by-montek-ahluwalia-and-jairam-ramesh-138314

Hossain, N., and D. Kalita. 2014. "Moral Economy in a Global Era: The Politics of Provisions During Contemporary Food Price Spikes,' *Journal of Peasant Studies* 14 (5):815–831.

Indian Institute of Population Studies. 2007. *National Family Health Survey 3. 2005–2006*. India Report. Ministry of Health and Family Welfare, Government of India.

Indian Institute of Population Studies. 2017. *National Family Health Survey 4. 2015–2016*. India Factsheet. Ministry of Health and Family Welfare, Government of India.

Jishnu, L. 2004. "Too Little on the Plate." *Businessworld*, 28 July www.businessworld.in/july2604/indepth03.asp

Khera, R. 2011. "Trends in the Diversion of Grain From the Public Distribution System." *Economic and Political Weekly* 46 (2):106–114.

Khera, R. 2013. *Democratic Politics and Legal Rights: Employment Guarantee and Food Security in India*. New Delhi: Institute of Economic Growth.

Kothari, R. 1984. "The Non Party Political Process." *Economic and Political Weekly* 19 (5):216–224.

Majumdar, B. 2007. "Food Riots Expose How Corruption Hurts India's Poor." *Reuters*, Friday October 12. http://in.reuters.com/article/idINIndia-29970920071012

McAdam, D., S. Tarrow, and C. Tilly. 2001. *Dynamics of Contention*. Cambridge: Cambridge University Press.

Omvedt, G. 1994. "Peasants, Dalits and Women: Democracy and India's New Social Movements." *Journal of Contemporary Asia* 24 (1):35–48.

Pande, S. 2014. "The Right to Know the Right to Live: Grassroots Struggles for the Right to Information and Work in India." PhD Thesis, Brighton: Institute of Development Studies.

Patnaik, B., and D. Sinha. 2016. "The Story of Paddy Procurement in Chattisgarh." In *Undernutrition and Public Policy in India: Investing in the Future*, edited by S. Desai, L. Haddad, D. Chopra, and A. Thorat. London: Routledge.

Programme Evaluation Organisation. 2005. *Performance Evaluation of Targeted Public Distribution System (TPDS), Report No.189*. New Delhi: Programme Evaluation Organisation, Planning Commission, Government of India. Accessed 13 January 2015 http://planningcommission.gov.in/reports/peoreport/peo/peo_tpdsmarch05.pdf

Sinha, S., B. Patnaik, V. Raaj, S. Bhattacharya, and A. Joshi. 2014. *Popular Actions, State Reactions: The Moral and Political Economy of Food in India*. Brighton: Institute of Development Studies.

Svedberg, P. 2008. "Why Malnutrition in Shining India Persists." 4th Annual Conference on Economic Growth and Development, New Delhi.

Swaminathan, M. 2000. *Weakening Welfare: The Public Distribution of Food in India*. New Delhi: Left Word.

von Grebmer, K., A. Saltzman, E. Birol, D. Wiesmann, N. Prasai, S. Yin, Y. Yohannes, N. K. Kabra, and C. A. Ittyerah. 1992. *The Public Distribution System in India*. New Delhi: Eastern Books.

7 The constitution lies to us!

Food protests in Kenya, 2007–2012

Celestine Nyamu Musembi and Patta Scott-Villiers

Introduction

> The constitution sometimes lies to us because they haven't told us where we will find that food that is our right. . . . You will die of hunger if you overstate the importance of that constitution.[1]

Labour Day, 2009: from the densely inhabited informal settlements of Nairobi, streams of people marched towards the city stadium. Their banners and slogans protested the cost of the staple maize meal, *unga*. '*Unga* 30 bob!' was the cry, harking back to a time when the price of a kilo of this most important of Kenyan foods was protected by government and rose at a rate of one shilling a year over a period of 24 years. Numerous localised riots also broke out in Nairobi's informal settlements at food distribution depots. Meanwhile, market stallholders at Ikutha, a marginal farming area 225 km from the capital, described noisy episodes outside grain stores and administration offices. These sporadic and localised protests were widely debated on local language radio, but received little or no national media attention.

In 2010, the Republic of Kenya enacted a new constitution that brought into law a range of progressive economic and social rights including the legal entitlement of its citizens 'to be free from hunger, and to have adequate food of acceptable quality' (Republic of Kenya 2010: 79). Yet the following year, in 2011, there were more protests as food prices peaked again and people on low incomes both cut down on the quality and quantity of the food they ate and stepped up their efforts to find more cash paying work (Scott-Villiers *et al.* 2016). Hunger was (and still is) widespread in Kenya. FAO statistics indicate that 10 million people – 23 per cent of the population – were undernourished in the years 2010–2012.[2] Not having enough to eat is commonplace and protests about dearth are usually localised skirmishes in marginal areas, raising little political attention. Factors rooted in Kenya's history and political economy have dampened citizen expectations of the state, thwarted popular mobilisation and allowed for persistently weak state responses, a process that this chapter will explore.

Larger protests do occur at certain junctures, and this was the case in the years 2009 and 2011 in particular. They were triggered by unprecedented rises in food prices and the cost of living, and inadequacy in state response. They were made legitimate in the minds of protesters and for many in government by a belief in state responsibility for provisioning at times of dearth. They had greatest effect (as far as they had any effect at all) when political dynamics allowed citizen pressure to be most acutely felt by politicians. The protest events thus helped reveal the makeup of the moral and political economy of low-income Kenyans in relation to food and other basic provisioning. We understand the idea of the moral economy to be the rationale for action by poor people, based on popular consensus as to traditional rights and customs at times of dearth (Thompson 1971: 79). In Kenya, it was brought forth in the shouts for price control and fair distribution of relief during the protest events, and in people's post-event explanations of their grievances and hopes. In this chapter, we use this notion of the moral economy to explore the politics of provisions in Kenya, tying the legitimising ideas of the moral economy into the opportunities offered by political dynamics and networks of solidarity and reciprocity. The result is a political bargaining process between the state and the masses which has substantial effects on state formation (Bohstedt 2016: 1035). The nature, timing and effect of the food protests in Kenya revealed the political opportunities to which the events responded, the networks of solidarity and fracture among people most affected by the price rises, and the degree to which historically-consolidated reciprocity (however weak) between poor citizens and state authorities still existed. This exploration of the politics of provisions allows insight into the means by which bargaining in favour of a contemporary moral economy might develop in the future.

We attempt here to show the field of struggle, by examining how local participants explained the contentious events in which they were involved, and analysing how elites and people on low incomes interpreted official responses to protest. It is based on a study carried out in 2013, which combined analysis of media reporting, policy, academic and legal documents with focus group discussions and validation meetings with low income residents in two sites, one urban and the other rural, along with interviews with government officials and right to food activists (Musembi and Scott-Villiers 2015).[3] Mathare, one of Nairobi's informal settlements, was a significant site of organising for the *unga* (maize meal) price protests that followed spikes in prices in 2008 and 2011. Ikutha market, the rural site in Kitui County in the southeastern region, is among the areas in the country that are considered chronically food insecure,[4] yet which has not been the focus of government attention due to its low political importance.

The triggers

> A right to food means you are absolutely free to eat food if and when you can find it [Ni kuya wawona.][5]

Over the years 2007–2012, the price of maize – Kenya's main staple grain – rose, fell and rose again in a series of spikes, each of which left the price of this basic food higher than it had been before the spike (Figure 7.1). While the timing would suggest that the price volatility was a reflection of the global food, fuel and financial crises, closer examination shows that food prices in Kenya rose

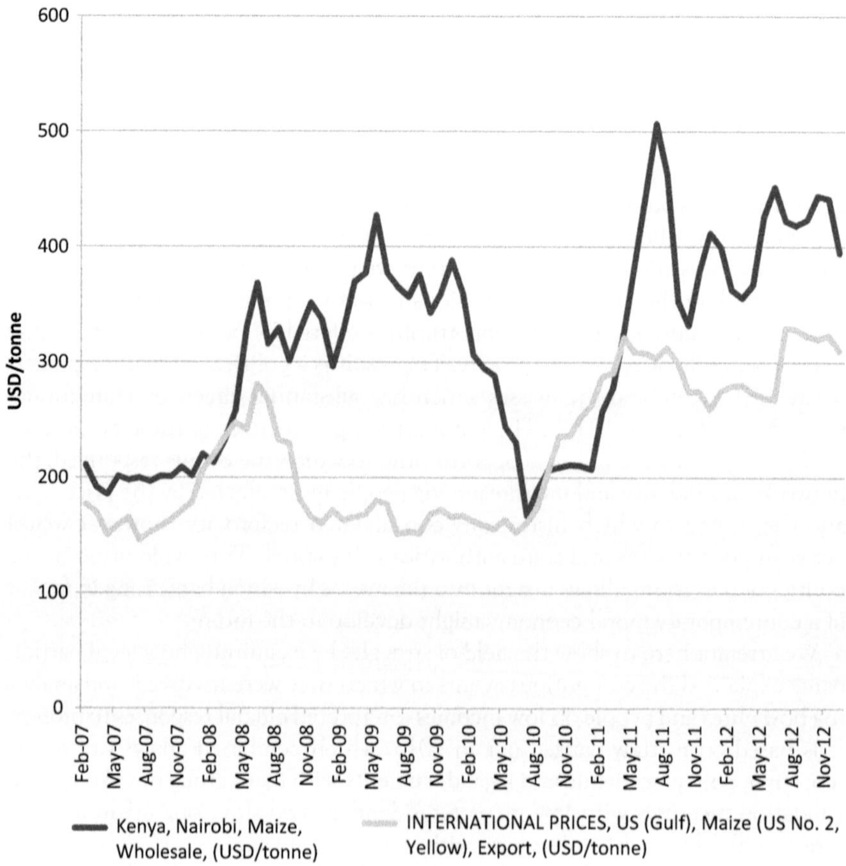

Figure 7.1 World and Kenya maize prices, 2007–2012

Source: FAO GIEWS, www.fao.org/giews/pricetool/ – accessed 28 February 2017

more sharply and stayed higher for longer.[6] Over the five-year period, though food prices responded to both local and international conditions, it was local conditions (production shortfalls, high import tariffs, dubious market interventions and failures of social protection) that kept prices rising. It is of these local conditions that poor people were speaking when they protested.

Costs of grain and oil on world markets began to rise rapidly in late 2007, while poor rains across the East African Community free trade area raised prices in regional and local markets. Between the end of December 2007 and mid-February 2008, unprecedented post-election violence in Kenya caused mass displacement from central maize-producing areas and substantial losses in food production (Short, Mulinge, and Witwer 2013: 11, Fengler and Kiringai 2009: 7). The overall effect was a deficit of some 1 million tonnes and a trebling in retail maize prices. People living on low incomes in rural Ikutha and urban Mathare explained how the price of not only maize meal, but also wheat flour, rice, cooking oil, beans, vegetables and other basics soared, as did the costs of rent, agricultural supplies and transport.

Kenya's formal mechanism for reducing volatility in grain prices, the Strategic Grain Reserve, managed by the National Cereals and Produce Board (NCPB) has existed for several decades, its origin in the paternalist authoritarian government's view that the prices of staple foods should be kept steady. When prices rose in 2008, even thought the country's economy was now operating under a 'free-market' regime, the mandate of the NCPB was unchanged: to release stored grain onto the market, as it had done in previous instances of shortage.[7] However, the stores were almost empty (Höffler and Ochieng 2009: 29) and there was no direct government action to mitigate the price shocks for almost a year.[8] In November 2008 the government announced that NCPB would import maize using aid from the Japanese government and sell it to millers at below wholesale market prices, on the basis that the millers would then make the milled maize flour available on the retail market at a low price.[9] Despite this initiative, the consumer price of maize flour remained stubbornly high (Höffler and Ochieng 2009: 29, Jayne and Tschirley 2010: 8, Short, Mulinge, and Witwer 2013: 11, Tschirley and Jayne 2010). At least half of the maize allocations of the scheme's first phase went to only 10 companies and a quarter was issued to brokers, who sold on to the millers at a mark-up (Fengler and Kiringai 2010: 7). In January 2009, local media reported the failure as corruption, including alleging that the duty free imports had been re-exported to Sudan (Ariga, Jayne, and Njukia 2010: 16). In February 2009, the scheme was abandoned. Public losses were estimated at 23.4 billion Kenya shillings (Fengler and Kiringai 2009: 13) – US $310 million – and the scheme joined the lexicon of Kenyan scams in the media, on social media and the word on the street in the informal settlements.[10]

Kenya's governance has long been characterised by an intricate dovetailing of economic and political power, sharply manifest in the food and agriculture sector (Leys 1975). A post-colonial elite inherited the colonial practice of subsidising large-scale farmers, while privilege quickly congealed into entitlement, secured by political clout. The favouring of maize-surplus areas both in pricing policy and production subsidies bears witness to this. Cereals board subsidies, for instance, have been shown to benefit a small number of large growers through raising the price of maize by as much as 20 per cent (Jayne, Myers, and Nyoro 2008). The maize-surplus areas of the Rift Valley and Western Kenya happen to be vote-rich areas, so that any suggestion of radical reform of the NCPB raises political temperatures. By contrast, far-flung Ikutha can be safely ignored, and repeated years of drought and chronic food shortage can fail to register on the political radar screen. Experience has also shown that rumblings in urban slums can be 'waited out', repressed or appeased with hurriedly crafted relief programmes that do not need to last.

> The people who grow maize for subsistence are not the ones who are subsidised by the cereals board. The maize sector is held hostage by vested interests. The whole thing subsidises 2 per cent of farmers from the North Rift

> and Kitale who are rich! They sit with us in cabinet and they tell us not to interfere with maize prices.[11]

Those most affected by the soaring prices were already spending as much as half their income on food from the market. Even small farmers suffered, since most were net buyers of grain. In 2008–2009 and 2011 we learned that it was common for families and especially for women to be eating only one meal a day. They ate smaller portions, reduced the variety of food to mainly *ugali* (cooked maize meal) and kale, and ate less diverse diets (Hossain and Green 2011, Lubaale 2013, Gatimu 2014). Many in Mathare began to rely on eating small portions of ready-cooked roadside foods rather than home-cooked meals, cutting down on fuel costs and saving time. In Ikutha the *kadogo* economy, or the re-packaging of basic food items such as flour, maize, sugar and cooking oil into ever-smaller portions to be bought by people barely scraping by as prices rose daily, was cited as a recent significant change brought about by rising food prices.

In the face of looming destitution, many people took often-irreversible decisions to change their diets, their work and the places where they lived (Scott-Villiers *et al.* 2016). Focus group discussants in Ikutha and Mathare explained which elements of the government's crisis response they felt breached the unwritten compact of a moral economy. While willing to double their own efforts to earn cash, and tighten their belts in terms of expenditures and food consumption, they directed their contempt towards inadequate and corrupt practices of price regulation, social protection and relief. They considered the failures of government price stabilisation and the unreliable and unfair distribution of relief commodities to be utterly illegitimate. They explained that these failures, which the government could have prevented, had exacerbated the negative effects of global market turbulence and local production shortfalls. People were exasperated, though not surprised, by the degree of plundering and blundering on the part of their government and its cronies. In 2011, one *matatu* (public transport) tout went so far as to say: 'it's high time Kenya went the Egypt way. We need a leadership change.'[12]

Chronicle of food-related protest

> A right to food means that the price of food should be lowered by the government so that people eat well.[13]

Demonstrations, court cases, petitions and riots

The period witnessed cost-of-living-related strikes by public sector workers including health workers, teachers and lecturers[14] and violent riots by informal traders evicted from their pavement pitches by armed police.[15] Some of the most notable protests reported in national media related specifically to the cost of maize meal. These non-violent protests came to be identified as the '*Unga* Revolution,' led primarily by the poor people's social movement *Bunge la Mwananchi*.[16] Away from the eyes of the national media, local chapters of Bunge la Mwananchi had been hard at work organising protests over the rising cost of living since 2005.[17] They were doing this both in their vibrant weekly outdoor debates in Nairobi's Jeevanjee Gardens, a public park, and in campaigns within low-income neighbourhoods (Otieno *et al.* 2016). Through weekly forums in town centres, volunteers presented citizens with illustrations of the volatility and steep rise of *unga* prices since 2002. They compared the current situation to the relative stability in food prices during the era of Kenya's one-party state under Daniel Arap Moi and his Kenya African National Union (KANU) party. They presented a moral economy argument: that even a dictatorship kept its bargain with the poor to control the prices of staple commodities.[18] This counter-intuitive proposition achieved the desired effect of stirring up public outrage and collective action to protest.[19]

Although the name *Unga* Revolution has now become something of an institution, it first appeared in the national media as the chant of a gathering of workers heckling the labour minister reading the president's speech at the official Labour Day celebrations on 1 May 2009.[20] The minister was forced to cut short and eventually abandon the reading, bringing the celebrations to a premature end. The *Unga* drama was repeated during the Labour Day Celebrations in 2011.[21] Food price protesters had by then perfected the strategy of stealing publicity at national celebrations. This strategy was successfully employed during the Jamhuri Day celebrations in December 2008, most Labour Days since 2009, and Madaraka Day[22] in 2011. The activists consider their action a legitimate reclaiming of national forums that have been turned into exclusive spaces for the political elite. Alongside this unsettling of official functions, activists also maintained a two-decades-old tradition of observing the unofficial (anti-official) Saba Saba Day (7 July), which commemorates a brutal crackdown on proponents of multiparty democracy in 1990. The price of food was the theme of the Saba Saba rallies in 2011 and 2013. Speakers at both events exploited to full effect the historical link between the securing of 'political freedom' in the 1990s and the pursuit of the 'social and economic freedoms' promised by the 2010 constitution.[23]

Theatrical street demonstrations, particularly in the Nairobi city centre, attracted much media attention, especially when accompanied by confrontation with police. On 31 May 2011, *Unga* Revolution protestors took over Harambee Avenue, the location of the Office of the President and the then Office of the Prime Minister, temporarily renaming it "*Unga* Avenue". The media were delighted by this, but showed no interest in the intense behind-the-scenes mobilisation in low-income neighbourhoods that preceded each street event. It is on account of this less visible mobilisation that we encountered some very well-informed analysis of food politics in poor neighbourhoods such as Mathare.

The Consumers' Federation of Kenya (COFEK)[24] joined Bunge la Mwananchi in the protests in 2011 when the government announced a steep increase in the price of fuel that would affect food prices,[25] and on 30 May COFEK filed a lawsuit against the government offices concerned with regulation of the energy sector.[26] The court process in Kenya has historically been used as a site for public protest, and COFEK wanted the court to issue a declaration to the effect that the government agencies' failure to stabilise and bring down fuel prices had the effect of raising the cost of food, thereby breaching Article 43 of the constitution. On the day of the hearing, the court building and grounds were packed with people wielding placards and the event was heavily covered in the media. However, COFEK's petition was poorly substantiated: no impact data was presented; no demonstration of failure of the measures the government had taken; and the petition collapsed.[27] The one documented incident of court action in protest at food price hikes proved a wasted opportunity. It was rumoured that the COFEK leadership and legal team were targeted and divided with offers of deals and pay-offs, hence the poor showing in the handling of the case.

The next time food-related protest received national media attention was June 2013.[28] The protests were against a proposed revision of the Value Added Tax (VAT) Act to apply the full rate of VAT (16 per cent) on products (including basic food commodities) which had previously been either zero-rated or exempted.[29] Packaged *unga*, which had enjoyed zero-rated status since 2003–2004, was also included, earning the proposal the tag '*unga tax*' on the streets and in the national media. A coalition of organisations led by the National Taxpayers Association held protest rallies for three days (24–26 June 2013) following the budget speech in which the proposed measure was announced, and then held a protest march to the Treasury on 2 July 2013.

Petitions were not uncommon in association with these demonstrations and marches. They involved the background work of getting enough signatures, then a high-profile march to the relevant office to present the petition. The *unga* tax protests, for instance, had already collected more than 10,000 signatures on their petition a year earlier. In most instances, police blocked the protesters from accessing the official to whom they intended to present the petition. In a rare gesture, one of the *Unga* protests (31 May 2011) saw the prime minister come out onto the street to personally receive the petition.[30] In all probability, such a response was only made possible by the internal rivalry of a coalition government. However, the prime minister's appearance did not translate into any lasting food policy reform. He requested 14 days to exercise his 'right of reply', but he never did get back to the people. It was an instance of bad faith in relation to the moral economy that community activists were still angry about two years later. They speculated that his posture, identifying with the troubles of the poor, was merely a cynical exercise in media attention. Being the party leader of the weaker coalition partner, it is likely that the PM was unable to mobilize an official response. Despite their initial jubilation at his appearance, activists soon realised that there was little real hope of food policy reforms.

Besides the more organised mobilisations, there were also many sporadic and localised riots. We heard about these in our field research in both Mathare and Ikutha. In Mathare spontaneous skirmishes were witnessed repeatedly at the depots where the urban poor tried to buy the government's much-touted subsidised maize meal. They were also witnessed during food distribution to Mathare's internally displaced persons, who were still homeless five years after the 2007–2008 post-election violence. The skirmishes, which went unreported in national media, were triggered by alleged manipulation of the beneficiaries' list by local administrators.[31]

In Ikutha we became aware of at least three protests against local administrators in the distribution of relief food. The first two were directed at a district commissioner and a district officer for their personal involvement in hoarding and then re-selling relief food. In the first incident the protesters, with the backing of a local councillor, succeeded in preventing a truck loaded with relief food from leaving the district commissioner's compound, and forcing instant distribution of food stockpiled in a government warehouse.[32] The second incident took place in Kituti, one of the outlying market centres. Local young men had been employed as casual labourers to offload a large consignment of relief food. The lorry then left empty, and returned with a second consignment. While they were still waiting for the order to off-load, the lorry started driving out of the district office premises. The young men blew the whistle, but by the time a crowd gathered it was too late. However, they vowed that they would not leave, and the agitated crowd kept growing, until every last grain from the first consignment had been distributed. The third was even more localised, directed at an assistant chief for allegedly stashing relief food away in a store room in his office. His defence was that he was storing it temporarily for some beneficiaries who had not attended the public distribution. 'When we found a hammer and he saw that we were serious, he agreed to unlock the door.'[33] In the meantime, the protesters had made calls to their elected councillor and to a local language radio station that captured the drama as they emptied out the store and distributed the food.

State response to the protests

> The PM promised affordable unga, but that only lasted a few days. Is there anything that moves the government to take more serious action?[34]

The immediate response from the state varied from repression to appeasement, including shows of solidarity by individual politicians. Repression took the form of blocking protesters from accessing government offices, beatings and intimidation of protest leaders.[35] Blocking access led to violent confrontation between the police and protesters, as was the case with the *Unga* protests of 2011. These violent confrontations were then followed by arrests but seldom by any formal charges. Denial of (or delay in offering) medical treatment to those injured by police was also common. Threats to movement leaders were also part of the repressive response.[36] One of the community activists in Mathare found herself confronted in her police cell by the three most senior police officers in the district, all from her ethnic group, deriding her for having been brainwashed by an 'enemy' tribe: 'Don't you know that "we" are now in power? What are you protesting about?'[37]

Appeasement was exemplified by token responses such as a miniscule reduction in excise duties on kerosene and diesel, leaving the hefty fuel levy untouched (Njagih 2011), or the short-lived availing of a small amount of subsidised *unga* through market outlets in urban areas. Maize meal was specially packaged and delivered to designated depots in poor neighbourhoods between December 2008 and March 2009, but people in Mathare claimed that almost no one was able to get access to it.[38]

There were several instances of individual politicians ostensibly lending support or expressing agreement with the protests. The *unga* campaign could boast two presidential pledges to take measures to make food affordable, but these were characteristically not backed by anything in written actionable form.[39] Then there was Mike 'Sonko' Mbuvi, the maverick MP for Makadara (a Nairobi constituency) at the time of the 2011 protests, who tried to hijack the publicity of the *unga* protests of 31 May 2011 as part of his successful bid for election as Nairobi Senator.[40]

Only two instances drew an institutionalised response. The first related to the *unga* tax. A number of MPs had already opted in late 2012 to side with the *unga* tax protesters.[41] A looming general election in March 2013 might have had something to do with this populist gesture. The *unga* tax protest continued to attract MPs' support even after the general elections, notably MPs from the Jubilee coalition of parties that got into power, since lowering food prices had been one of their campaign promises.[42] It would be fair to say that this victory was a precarious one, because future re-classification of goods and services for VAT purposes is always a possibility, and the pressure of revenue collection targets could see the debate re-opened.

The second instance concerned the enactment of the Price Control (Essential Goods) Act in 2011.[43] In June 2010, in an apparently hasty response to the various cost-of-living protests in 2008 and 2009, parliament debated and passed a bill bringing several essential commodities under a price control regime. Economists described the bill as ill-thought-through.[44] The business community successfully pressured the president into withholding assent to the bill, citing Kenya's obligations under international and regional trade treaties. In June 2011, a parliamentary select committee on the cost of living was set up, whose work eventually led to a revised version of the bill. The president gave his assent and the bill was enacted into law in September 2011. However, this law represented a hollow victory. The Act is very brief. It gives power to the responsible minister to declare, through a notice in the official Gazette, any goods to be essential commodities and thereby set a maximum price 'in consultation with the industry'. Offences and penalties are provided for contravention of the law, but no institutional mechanism is legislated for its enforcement. Rather, the minister is simply given powers to enact regulations 'for the better carrying out of the intent and purposes of this Act'. It is not clear which ministerial docket has the mandate to implement this law: is it under trade/commerce, or is it under social protection? What if the minister fails to make the regulations? As is often the case, what started out as a parliamentary initiative ended up in the control of the executive and the business sector. Activists, whose power in these realms is weaker than their influence with parliamentarians, did not consider it worthwhile to engage.

In 2007, before the explosion of the food and political crisis of 2008, the government had made proposals to update the existing food and nutrition policy dating from the 1980s. The draft document departed from prior emphasis on maize self-sufficiency to more comprehensive food access, diversity and nutritional status (Republic of Kenya 2007). It recognised that high staple prices only favoured producers of surplus, while hurting urban consumers and the great proportion of rural producers who were net buyers and it proposed lifting import duties on grains and removing restrictions on cross-border trade (Ariga and Jayne 2010: 227). The revision was not implemented, however. A new set of policies was drawn up and adopted by the cabinet in 2011. The Food and Nutrition Security Policy of 2011 begins with the constitutional right to sufficient and nutritious food. It stipulates increased funding to the food and agriculture sectors, mandates increasing the strategic grain reserve and building the infrastructure to encourage local supplies into local markets and between surplus and deficit areas. It is silent on the much-debated problems of unequal access to subsidies and facilities, and on the inefficiencies in the strategic grain reserve and NCPB purchase and sales (Republic of Kenya 2011a: 13). On the achievement of adequate nutrition, the policy notes children's constitutional right to proper nutrition and lays out government responsibility for provision of supplements for new mothers, infants, adolescents, vulnerable adults and older people. On relief and welfare, it offers continued food relief in rapid-onset emergencies, cash transfers including cash for work (building assets like roads and dams) and cash safety nets for people who are chronically poor. This area is then covered in more detail in the National Social Protection Policy (Republic of Kenya 2011b), but neither document tackles the limitations of the current relief and welfare programmes in terms of cost, inadequate coverage or corruption. A recent review of social protection in Kenya reports that from 2005 to 2010 social protection expenditure rose from KES 33.4 billion (USD $371 million) to KES 57.1 billion, or 2.28 per cent of GDP (Republic of Kenya 2012: vi). The bulk of this involved spending on contributory programmes for public and private sector employees, including the civil service pension. External donors picked up most of the tab for the remaining part, providing 71 per cent of the costs of relief and hunger safety net (*ibid.*: vii-viii).

Government response to the food crisis and to the protests was roundly criticised among people on low incomes, though they recognised a few improvements to some social welfare programmes and the victory of the VAT exemption struggle. Rights activists in Nairobi who have researched food shortages have come to the conclusion that inadequate grain reserves and subsidies,[45] combined with policies of support to large local producers and *ad hoc* waivers of import duties on staple grains have facilitated the commoditisation of food. One thing is clear: there is no lack of state capacity to create a system for preventing malnutrition. Kenya has an acclaimed system for famine early warning and response, albeit one alleged to be partly driven by short-term private gain. Nonetheless, it has worked to keep famine-related casualties low, and fatalities almost eliminated (World Food Programme 2011). According to the critics' view, the reason why government departments invested in a famine relief response mechanism but not in a long-term mechanism to deal with chronic hunger relates not to blunder, but to plunder: 'there is no interest in long-term engagement because long-term engagement does not make people rich'.[46] In short, activists believe that there is little interest in establishing institutions that work for the long term because 'we have a mafia that thrives on crisis.'[47]

The struggle between the moral and political economies

Based on our discussion, we can now lay out the grounds of the struggle that constitutes the politics of provisions in Kenya. Bohstedt (2016) emphasises two aspects that help in the analysis: horizontal solidarities and vertical reciprocities or challenges. As we have seen previously, government's response to the price hikes that led millions of families to cut down on food for significant periods of time was grossly inadequate. We saw both blunder and plunder in the management of subsidies, tariffs and relief. We saw limited reciprocity in the matter of the *unga* tax and tokenism in the matter of the Price Control (Essential Goods) Act in 2011. The political power of the masses, activists and the press in the moments of protest did not do much to change the overall trajectory of the political economy, channelled as it was through well-worn courses of appeasement and repression.

The moral economy of the poor

Reading about and listening to the views expressed by the *Unga* protesters, one comes away with the impression that, though often ironic, there is a sense of citizen entitlement and government obligation with respect to food and basic provisioning that builds on an era of paternal governance that preceded the current era of free trade:

> They cannot eat when we are not eating.[48]
>
> We wish to remind Mr Rotich [Cabinet Secretary for Finance] that he has misled the president. We are disappointed that even after coming from a humble background, he has failed the moral test by turning his back on the poor.[49]

The talk that established a sense of solidarity between ordinary market-goers, grassroots protestors, COFEK, non-governmental organisations such as the National Taxpayers Association and the media was of entitlement to some means of basic provisioning and/or to protest their lack. These sentiments can be said to represent the moral economy around food since they are legitimated beyond the protesting groups:

> By the notion of legitimation I mean that the men and women in the crowd were informed by the belief that they were defending traditional rights or customs; and, in general, that they were supported by the wider consensus of the community. On occasion this popular consensus was endorsed by some measure of licence afforded by the authorities.
>
> (Thompson 1971: 78)

Many of the views expressed in the focus group discussions and in media coverage undoubtedly suggest a degree of wider community consensus on the existence of citizen entitlement and government obligation, in particular in times of food crisis. However, at the grassroots level, there was also a strong sentiment that at a specific historical moment such a moral bargain had ruptured. That moment was identified as the imposition of *soko huru* (free market) under the Moi regime in the early 1990s, when price control was abolished (under the Structural Adjustment Programmes).[50]

After civil society pushed through economic and social rights in 2010, protesters made explicit reference to the constitutional right to food to establish an ethical basis for the government's obligation to protect its citizens from the effects of the food price hikes. They were calling for a right to fair governance of the markets and administrative systems supplying the basic essentials, notably food, and, at times of dearth, they called for subsidy. In the focus group discussions, we sought to understand how the idea of a right to food was viewed, and whether people might conceive of it in ways that were likely to spur political mobilisation. The findings revealed an ambivalence reflecting the struggle between the political and moral economies. A restrictive view interpreted the right to food simply as direct provision of free food by the government when and where need was greatest. This view limited the right to delivery of aid in situations of extreme destitution through a system vulnerable to graft. The category of rights-holders was narrowed down to those seen as 'vulnerable' or unable to feed themselves due to circumstances beyond their control. Viewed in this manner, the right to food did little to catalyse broad-based mobilisation.

Many others had a more expansive view, recognising that a right to food had to mean more than occasional free food. Here the emphasis was on creating an environment that would enable access to food and facilitating the ability to feed oneself. Thus, government measures such as regulation of food traders, stabilisation of prices, promotion of employment and investment in smallholder production was seen as being at the core of what a government must do in order to fulfil its obligation under a right to food. Taken together with other views that expressed outright cynicism about the utility of a right to food, these ambivalent perceptions explain why a rallying call for the right to food did not become a galvanising force after its adoption in the 2010 Constitution. The contemporary moral economy was at a low ebb, affected as it was by 20 years of free market and 'it's our turn to eat' graft.

The paternal aspect

> The price of maize meal and other essential commodities should be reduced to cushion the poor against the high cost of living.[51]

Did the government, by E.P. Thompson's criterion, endorse a paternalist view of the moral economy articulated in the protests and by many in the research sites? There were instances in which officials felt compelled to concede (or at least not dispute the claim) that the government had an obligation to cushion the poor against the high cost of living and that there was a political advantage to making such claims. Examples include MPs' support of the *unga* tax protest. The parliament's minority leader was quoted in the press as having said that the poor must be cushioned from the extra burden that would be imposed by VAT on food items, adding: 'If any measures are brought here through VAT that hurts the poor man we shall reject them. . . . We don't want those who live in poverty to be taxed harshly.'[52] Beyond the politicians, officials from the Kenyan Revenue Authority and the Treasury were careful during the VAT debate not to appear to confront head-on the notion of obligation to the poor. Rather, they argued that the existing regime had become captive to unscrupulous merchants adept at manipulating the inefficient refunds system, implying that for the poor to benefit there was a need for reform.[53]

Within government more generally, the reception of economic and social rights had been lukewarm, due to difficulties and costs of implementation and contradictions with the logic of market-based growth. With the exception of the social protection sector,[54] the message we got from government – especially the Treasury – was that rights had no bearing on the direction of policy.[55] Activists and champions within the government felt that the constitution was not deeply rooted in the day-to-day interaction between citizens and officials or politicians. Informal rules and ethnically defined identities and modes of engagement eclipsed the formal system archived in the constitution (Berman, Cottrell, and Ghai 2009).

If political leaders conceded the protesters' claims to a bare bones moral economy around food, what should we make of the violent repression of protest, or the tokenistic populist appeasement? A pessimistic interpretation is that the conviction did not run deep; that some politicians and officials only needed to give a nod in the direction of the moral economy asserted by the protesters for political expedience. A more optimistic reading is that the struggle between the moral and political economies went on inside government as much as it did on the ground: there are some within government who do have a conviction about a moral economy around food and basic provisioning, who repudiate both plunder and blunder. Others believe in a harsher market-based approach involving minimum levels of relief or price control.

The notion that the rich and powerful have obligations towards the poor majority has deep historical roots predating colonialism. However, the articulation of these obligations in post-colonial governance has taken on an increasingly materialistic and sectarian tone. Democracy in Kenya is largely clientelistic, and although considerable power was devolved in 2013 to 47 counties in the hope of spreading benefits more equitably, the tribalistic mode of governance still persists (Boone 2014; Cheeseman, Lynch, and Willis 2016; D'Arcy and Cornell 2016). It is usual for campaigning politicians to distribute substantial amounts of cash to potential voters, from their own and allied tribes. Cash handouts in election years dampened complaints about high food prices and unemployment. Aware that this was not quite in line with democratic rhetoric, people in our focus group discussions spoke in the third person about accepting campaign cash. In the extreme, it provides a justification for corrupt and unaccountable leadership: as long as 'one of our own' is in power there will be distribution of wealth to 'us'. The moral economy is fragmented in a way that has little of Bohstedt's (2016) horizontal solidarities for creating political punch. It is not necessarily a state of affairs that can last, since the vertical reciprocities also appear weak: many of those we spoke to admitted that once elected, political leaders showed little concern for the hungry in their own ethnic groups. They disappeared to parliament or the county capital and were not seen until the next election.

The politics of provisions in Kenya

The picture that emerged from our study is one of overall disillusionment with the political class. Would it be naïvely optimistic to hope that this collective disillusionment can be tapped and mobilised to transform the culture of political engagement? One Ikutha ironist's quip (after our long discussion on the failures of the political class) would suggest so – 'But mind you, if they were to come here now, we would dance for them.'[56] Both rulers and ruled are performing to a script tacitly authored decades ago. Kenya's politics of provision thrives on cynicism, irony and spectacle.

It became clear early on in our study that mobilisation and engagement with government on social and economic rights had been episodic rather than sustained and had not achieved as much as it might have. Kenya cannot credibly claim to have a 'right to food' movement in the sense that India does, for instance (Chapter 6, this volume). Government repression and appeasement shows that the structure of existing political bargaining has undermined any efforts to turn the protests into a mass movement. This is not surprising, since food is a matter on which citizens and state have established some basic rules of engagement – as Bohstedt points out, 'food rioters are usually not attempting to change the political system, but rather to summon it to their needs' (Bohstedt 2016: 1038). Nonetheless a movement could shift the balance of the bargaining arrangements, so what is getting in the way? The political class is unshaken by short-lived 'branded T-shirt' types of protests, but 'we would quickly come up with something if Kibera[57] was burning.'[58] The absence of a sustained social movement means that the political class can get away with simply quelling noise.

We believe that a primary obstacle to more sustained pressure is to be found in the history of mobilisation for rights in Kenya and the role of intermediary rights groups in diverting direct reciprocal relations between citizens and the state. Explicit mobilisation employing the language of rights only came into its own in the struggle against the one-party state (late 1980s–1990s). It intensified in the ensuing clamour for a new constitution. The struggle against the one-party state focused on political competition and securing the civil and political freedoms that would support it. It failed to connect these macro-political struggles with grassroots concerns that intertwined political with *economic* and *social* repression.[59] The public face of the rights struggle was that of a professional elite – a few budding opposition politicians, lawyers and prominent outspoken clerics, who were perceived as unable to relate to day-to-day struggles for basics such as food (Berman, Cottrell, and Ghai 2009, Kanyinga 2004). These basics were pigeonholed as 'development concerns', the province of an entirely separate cluster of actors – community development organisations. It was these organisations, funded through international donors, who took on the task of provisioning: filling the gap left by the state's failure to deliver services, rather than pressing for state responsibility. Their paths did not often cross with those of the 'rights' organisations until the early 2000s (Nyamu-Musembi and Musyoki 2004, Okello 2004). The effects of this history are still evident. The *Unga* Revolution did not include in its ranks names such as the Kenyan Human Rights Commission or the International Commission of Jurists, both of them affluent NGOs outspoken on human rights issues since the 1990s.

A second factor is a rural-urban chasm, a fractured horizontal relation between people who may have common ground (they all suffered acutely from rises in the cost of food) but who had felt isolated from each other. The kind of mobilisation that attracted media attention and sought direct engagement with high-ranking state officials was mostly a Nairobi-based affair. With very few exceptions, the rural scene only recorded small-scale protests such as those we noted in Ikutha which were not covered by the media. National protest campaigns did not tap into the energy expressed in these widespread but localised protests and channel it into sustained broad-based mobilisation. To be fair, Bunge la Mwananchi, the chief protagonist in the *Unga* Revolution, has a structure with a grassroots reach that cuts across urban and rural. However, its leaders acknowledge that they have had more success in organising the urban poor, and their most visible campaigns have been the ones staged in the capital.[60] This is in contrast to India, where the 'right to food' movement draws most of its participation from a rural constituency across the various states (Sinha *et al.* 2014). The explanation must lie in the political opportunity structures of the rural areas, which limited the effectiveness of protest.[61]

Our review of media reports on food-related protests in the 2007–2012 period showed little evidence of media profiling any emergent right to food movement which might have helped define the terms of public discourse, a third factor in the failure of sustained public pressure. Media reports tended to report the confrontation with authorities, especially when these had partisan political undertones, mentioning only in passing the issue at the heart of the protest.[62] Such reportage is part of the structure of clientelist politics. There were no fact files on the campaign or media briefing packs to ensure accuracy and depth in reporting. We found no opinion pieces authored by the campaign's leaders to inform the public, influence opinion or offer an alternative to the state's narrative. Again, this is in contrast to India, where strategic alliances with the media have been central to sustaining the mobilisation, making it relevant to public discourse and impossible for the authorities to ignore (*ibid.*).

In the years after the 2008 and 2011 crises, the moral economy was strongly tempered by the free-market political economy, which emphasised the privatisation of misery in the name of resilience and economic growth. In the words of one policy maker, 'somehow people survived and life continued.'[63] The dovetailing of economic and political interests in Kenya's evolving governance system, coupled with the way in which electoral politics has been built on ethnicity, makes for an unfavourable political opportunity structure for movement organizing. A strong sense of personal economic responsibility, the chronic nature of hunger, the debilitating effects of poor diet and the experience of repeated disappointment in politics and donor-funded civil society has made patron-client arrangements more attractive than public collective fury. These constraints on reciprocal relations between the common people, intermediaries and rulers have played an important part in restricting the power of food protests in Kenya. Nonetheless, the part played by the *Unga* Revolution in the evolution of state institutions has been to demonstrate that, were clientelistic tribal politics to fail, future governments will face a formidable challenge from the people.

Notes

1 Roadside food seller, Mathare, June 2013.

2 FAO, The State of Food Insecurity in the World, www.fao.org/hunger/en/ accessed 06/01/2017.

3 Primary and secondary data collection for this study was a joint activity by the University of Nairobi, Kenya and the Institute of Development Studies, UK. The researchers included the authors, Bonface Omondi and Nicholas Benequista who led on the media study; and Ann Mbuthia, Claris Akinyi, Stephen Irungu and Jason Waweru in Mathare and Benjamin N. Kiema, Eirc Kitheka, Cynthia Muinde, Stephen K. Mulatya, Sharlet Martha Musingi, Orah Mwende, Malombe Mulinge and Elizabeth Ngumii in Ikutha who led on the field work in the respective sites. The study was part of a four-country project entitled 'Food riots and food rights: the moral and political economy of accountability for hunger', led by Naomi Hossain (IDS) with colleagues in Bangladesh, India, Mozambique, Kenya and the UK and funded by the Economic and Social Research Council / Department for International Development Joint Fund for Poverty Alleviation Research (Grant reference ES/J018317/1). We are grateful to participants in the national reference group meetings and with special thanks to Professor Yash Ghai CBE, Chairman of the Katiba Instiute; Cidi Otieno, Gacheke Gachihi, Rachel Mwikali and many others from Bunge la Mwananchi; James Oduor, Director of the Kenya National Drought Management Authority; the Hon. Mohamed Elmi, Member of Parliament of the Republic of Kenya; and the people of Mathare and Ikutha who gave up their time to take part in and contribute with such zest to our listening posts, interviews and focus group discussions. We would also like to thank our colleagues from Mozambique, India, Bangladesh and the UK for the excellent feedback and important conversations.

4 Low agricultural productivity and erratic rains have resulted in perennial food shortages in the district. (Republic of Kenya n.d.)

5 Traders near Ikutha Market, Kitui, July 2013.

6 Staple food prices in Kenya have generally not shown a long-run relationship with world markets (Minot 2011: 24). Though there is considerable disagreement as to why, there is a strong argument that high import tariffs accompanied by mismanagement of grain and fertiliser subsidy interventions is a plausible cause (Jayne, Myers and Nyoro 2008).

7 Jayne, Myers and Nyoro (2008) note, for instance, that NCPB released grain onto the market in the 1992–1993 drought and again in 1993–1994, thereby lowering prices.

8 The sorry state of the NCPB and its inability to respond was confirmed in interviews with government officials: Interview, National Drought Management Authority, 11 June 2013; Interview, Ministry of Special Programmes, 03 July 2013; Interview, Food Security and Early Warning Programme, Ministry of Agriculture, 22 July 2013.

9 The Japanese government provided 940 million yen in 2009 and 630 million yen in 2010, which was used to purchase 42,000 tonnes of non-GMO maize, 90 per cent of which was sold to millers through NCPB at below market prices. The market price at the time was KES 3,500 per bag of maize, and millers got it at KES 2,700 per bag. The remaining 10 per cent was distributed to areas of the country with the highest vulnerability, through Kenya Red Cross and Catholic Relief Services. Interview: Food Security and Early Warning Programme, Ministry of Agriculture, 22 July 2013.

10 The scandal led to the sacking of the entire NCPB board of directors and 17 senior managers in early 2009 (Jayne and Tschirley 2010: 9), but no one was prosecuted and no further information was given to parliament or the Kenyan public about the scheme.

11 Interview with Member of Parliament (name withheld), Nairobi, July 2013.

12 Comments of an unidentified protester during the 31 May 2011 protest in Nairobi city centre. Interview in Nairobi, July 2013.

13 Laundress, Mathare, Nairobi, June 2013.

14 See, for example, www.brookings.edu/blog/up-front/2012/10/18/labor-strikes-are-becoming-a-rising-concern-across-africa/ (accessed 31/01/17).

15 See, for example, Mombasa turned into battle zone as street traders riot, www.nation.co.ke/news/regional/1070-605950-ax5kmfz/index.html (accessed 31/01/17).

16 Bunge la Mwananchi [literally, the ordinary person's parliament] was founded in 1990. It started as a ragtag movement discussing inter-religious harmony between Christians and Muslims, following several incidents of conflict, then broadened out to constitutional reform. At a time when few dared to speak out, *Bunge* dared to identify its mission as providing a public platform where ordinary citizens could discuss constitutional reform to address social justice issues affecting them. Martin Shikuku, an outspoken critic of the Kenyatta and Moi governments and a veteran independence-era politician, was actively involved in its foundation. Consolidation of a leadership structure within the movement only came after 2002. There is a *bunge* [parliament] in every ward, some very active and others not so, culminating in a national *bunge*. The *Unga* Revolution drew its energy from *bunge* chapters in Nairobi's informal settlements.

17 Email communication from Cidi Otieno, former convenor, Bunge la Mwananchi, 24 January 2014.

18 Maize prices had been subject to tight control under President Kenyatta in the 1960s, and the policy continued under the Moi government, at least until 1993 when economic crisis and IMF pressure led to liberalisation. In 1981, after severe food shortages, the Moi administration introduced a national food policy (Republic of Kenya, Sessional Paper no. 4 of 1981) which emphasized national self-sufficiency in maize and made provision for adequate relief to people in need in instances of shortage: 'to ensure adequate supply of nutritionally balanced food in all parts of the country at all times'.

19 Interview with Cidi Otieno, Convenor, Bunge la Mwananchi, 3 June 2013.

20 See L. Barasa, 'The day Kenya workers' anger boiled over', Daily Nation, 1 May 2009, www.nation.co.ke/News/-/1056/593070/-/u6683d/-/index.html.

21 See A. Mutambo, 'Hunger-driven drama at workers' fete', *Daily Nation*, 1 May 2011, www.nation.co.ke/News/Hunger-driven-drama-at-workers-fete-/-/1056/1154518/-/50p3p0z/-/index.html.

22 Public holiday on 1 June marking the day in 1963 when Kenya achieved self-rule. Jamhuri day is observed on 12 December to mark the day in the same year when Kenya became a republic.

23 See D. Opiyo, 'Nairobi demo turns chaotic, 10 people arrested', Daily Nation, 7 July 2011, www.nation.co.ke/News/Nairobi-demo-turns-chaotic-10-people-arrested/-/1056/1196202/-/ve51d2z/-/index.html. Also aired on Citizen TV: www.youtube.com/watch?v=WKo-Y4Kj8WA; and on NTV: www.youtube.com/watch?v=5gWblabufAI. See also 'Lobby protests at bid to increase food prices', Daily Nation, 7 July 2013, www.nation.co.ke/News/Lobby-protests-at-bid-to-increase-food-prices/-/1056/1907958/-/wrtia4/-/index.html.

24 COFEK was founded in March 2010, basing its mandate on Article 46 of the constitution, which sets out consumers' rights. It is a membership organization, open to individuals who pay a joining fee of Kshs. 1000 (USD 10) and an annual membership fee of Kshs. 6,000 (USD 60). It is fair to say that COFEK is far from approximating a social movement. COFEK seems to operate as a channel for individual and collective grievances against public and private bodies seen as violating consumers' rights, writing petitions and widely publicized open letters to the relevant bodies and threatening court action. Apart from the 2011 lawsuit against the energy sector regulators, we found no evidence that COFEK has undertaken court action in response to public grievances. See www.cofek.co.ke.

25 See 'Consumer group calls for protest against rising prices', *Daily Nation*, 17 April 2011, www.nation.co.ke/News/Consumer-group-calls-for-protest-against-rising-prices/-/1056/1146110/-/eyqj0oz/-/index.html and 'Consumer Federation chief warned over food protests', *Daily Nation*, 26 April 2011, www.nation.co.ke/news/-/1056/1150446/-/lx233r/-/index.html. See also L. Barasa and M. Mwaniki, 'Kalonzo urges patience over cost of living', *Daily Nation*, 19 April 2011; and *Daily Nation* editorial, 'Don't expect endless patience as prices soar', 16 April 2011, www.nation.co.ke/oped/Editorial/Dont-expect-endless-patience-as-prices-soar-/-/440804/1145824/-/1feih7z/-/index.html.

26 Consumers' Federation of Kenya (COFEK) v. the Attorney-General, Minister for Energy, Minister for Finance, Energy Regulatory Commission and National Oil Corporation, Petition No.88 of 2011 (High Court, Milimani), available at www.kenyalaw.org.

27 The judge laments: 'Other than the statements set out above, the petitioner has placed nothing before the court that can assist the court in determining the alleged failures by the respondents.' The judge admonishes COFEK for not taking seriously and giving insufficient attention to the critical issue it was asking the court to decide. COFEK judgement, pp. 8, 10.

28 The gap is largely because 2012 was dominated by the campaigns for the national elections held in March 2013.

29 See K. Njoroge, 'As food prices rise, brace yourself for an Unga war', Daily Nation, 18 June 2012, www.nation.co.ke/Features/DN2/As-food-prices-rise-brace-yourself-for-an-Unga-war/-/957860/1886176/-/5k1s1uz/-/index.html; J. Anyanzwa and M. Kamau, 'Tax shocker in cabinet secretary Henry Rotich's budget in his bid to fund new rail system', EA Standard, 15 June 2013, www.standardmedia.co.ke/?articleID=2000086001&story_title=tax-shocker-in-cabinet-secretary-henry-rotich-s-budget-in-his-bid-to-fund-new-rail-system&pageNo=1; A. Mutambo, 'Activists to protest against "unga" taxes', Daily Nation, 16 June 2013, www.nation.co.ke/News/Activists-to-hold-Unga-tax-demo/-/1056/1885152/-/mklhgv/-/index.html and K. Njoroge, 'Groups set for rallies against VAT on food', Daily Nation, 24 June 2013, www.nation.co.ke/News/Groups-set-for-rallies-against-VAT-on-food/-/1056/1893408/-/e5qt3ez/-/index.html. Coverage on TV: NTV 25 June 2011, www.youtube.com/watch?v=o-m37zxGu5c.

30 See www.youtube.com/watch?v=GaTG8CgFKyU [Citizen TV, uploaded 19 June 2011, but it must have aired on 31 May, because of the message 'no Unga, no Madaraka Day']. See also coverage on NTV: www.youtube.com/watch?v=CZe5tX28cHk. See also Focus Group Discussion with community activists, Mathare, 10 June 2013.

31 Focus group discussion with internally displaced recipients of food aid, Mathare, 10 June 2013.
32 Focus group discussion with small-scale market traders, Ikutha, 29 June 2013; Interview with Hussein Mwandia, member of county assembly, Ikutha ward, Kitui, 26 July 2013.
33 Interview with Hussein Mwandia, member of county assembly, Ikutha ward, Kitui, 26 July 2013; interview with Francis Mutinda Sua, chief, Simisi location, Ikutha district, 26 June 2013.
34 Focus group discussion with young people, Mathare, 10 June 2013; interview with Cidi Otieno, Convenor, Bunge la Mwananchi, 3 June 2013. Interestingly, we were unable to find national media reports of this confrontation between Mike 'Sonko' (then a member of parliament) and the unga protesters.
35 Threats to movement leaders were also part of the repressive response. See 'Consumer Federation chief warned over food protests', Daily Nation, 26 April 2011, www.nation.co.ke/News/-/1056/1150446/-/10xunrnz/-/index.html; Focus group discussion with community activists, Mathare, 10 June 2013.
36 Focus group discussion with community activists, Mathare, 10 June 2013.
37 Focus group discussion with community activists, Mathare, 10 June 2013.
38 Focus group discussion with community activists, Mathare, 10 June 2013. Government officials acknowledged that this intervention was ill-conceived and hurriedly implemented. Interview, National Drought Management Authority, 11 June 2013.
39 See 'Kibaki pledges action on food prices', Daily Nation, 21 April 2011, www.nation.co.ke/news/Kibaki-pledges-action-on-food-prices/-/1056/1148994/-/mfgfus/-/index.html; F. Olick, 'We will lower cost of food, President Uhuru Kenyatta assures Kenyans', EA Standard, 21 June 2013, www.standardmedia.co.ke/?articleID=2000086431&story_title=Kenya-we-will-lower-cost-of-food-uhuru-assures-kenyans.
40 See Kenya Broadcasting Corporation (KBC) clip: www.youtube.com/watch?v=J64zl3vQpG4.
41 See A. Gitonga, 'MPs warn of food riots over VAT hike', EA Standard, 27 July 2012, www.standardmedia.co.ke/?articleID=2000062763&story_title=MPs-warn-of-food-riots-over-tax.
42 See F. Olick, 'We will lower cost of food, President Uhuru Kenyatta assures Kenyans', *The Standard*, 20 June 2013.
43 Price Control (Essential Goods) Act, No. 26 of 2011 (Chapter 113 of the Laws of Kenya), available at www.kenyalaw.org.
44 See, for example, 'KIPPRA: Debating the merits of price controls', TTI Stories of Influence, www.thinktankinitiative.org/sites/default/files/KIPPRA%20Debating%20the%20merits%20of%20price%20controls.pdf.
45 Underfunding of grain reserves, it seems, is not unique to Kenya, and neither is the history of corruption in managing them (Murphy 2009).
46 Activist, Nairobi, June 2014.
47 Activist, Nairobi, June 2014.
48 Comments of an unidentified protester during the 31 May 2011 protest. The phrase 'to eat' has a strong meaning in Kenyan political talk, since it implies plundering the state as well as having food. The Minister for Finance had just announced the government would stop using Mercedes-Benz cars and would instead assign Toyota cars in order to save on fuel costs, a downgrade that some ministers opposed. See video clips from Citizen TV and NTV respectively: www.youtube.com/watch?v=GaTG8CgFKyU and www.youtube.com/watch?v=CZe5tX28cHk.
49 Attributed to 12-point petition presented by COFEK-led protesters. See A. Mutambo, 'Groups stage demo against '*unga*' taxes', *Daily Nation*, 2 July 2013.
50 Focus Group Discussion with Roadside food-sellers, Mathare, 11 June 2013; Focus Group Discussion with Retailers, Ikutha, 27 June 2013; Focus Group Discussion with Butchers and kiosk operators, Mathare 11 June 2013.

51 Comment attributed to Professor Yash P. Ghai, former chair of the Constitution of Kenya Review Commission that presided over the process of enacting a new constitution, which culminated in the 2010 constitution. Professor Ghai was taking part in the fuel price protest led by COFEK on 19 April 2011. See 'Kalonzo urges patience over cost of living', Daily Nation, 19 April 2011, www.nation.co.ke/News/-/1056/1147378/-/10ycd5iz/-/index.html.
52 See 'Protests, strike threats, greet Uhuru Budget', *Daily Nation*, 14 June 2013.
53 Kenyan Revenue Authority, 'The Case for VAT Zero Rating to Achieve Social Policy and Distributive Objectives' (undated).
54 The social protection sector expanded coverage over the period of the study, in particular in favour of children, the elderly and pre-and post natal welfare for women. See, for example, Republic of Kenya (2012) Social Protection Sector Review June 2012, Ministry of State for Planning, National Development and Vision, Nairobi: Government of Republic of Kenya.
55 Personal communication with a member of the Commission for the Implementation of the Constitution, Nairobi, February 2013.
56 Focus group discussion with small-scale market traders, Ikutha, June 29, 2013.
57 Kibera is the largest informal settlement in Nairobi.
58 Interview with Member of Parliament (name withheld) (Nairobi, July 2013).
59 This is not unique to Kenya. The kinds of rights struggles that have appealed to the middle class in Africa have been about narrow political freedoms such as press freedom. See Mamdani (1989); Ndegwa (1996).
60 Interview with Cidi Otieno, Co-ordinator, Bunge la Mwananchi, 3 June 2013.
61 In 2013 Kenya devolved significant legislative, budgetary and administrative powers to 47 county governments, fundamentally changing political opportunity structures in the country. This can be expected to have new effects on the politics of provision in the country.
62 For instance, the 7 July 2011 (*Saba Saba*) attempt to re-occupy Harambee Avenue to protest the continuing rise of *unga* prices in spite of the government's promise of action was reported in the *Daily Nation* simply as 'Nairobi demo turns chaotic, 10 people arrested'. Four short sentences convey the purpose of the protest, the rest of the story describing the clash with police.
63 Interview, National Drought Management Authority, 11 June 2013.

Bibliography

Ariga, J., and T. S. Jayne. 2010. "Maize Trade and Marketing Policy Interventions in Kenya." In *Food Security in Africa: Market and Trade Policy for Staple Grains in Eastern and Southern Africa*, edited by Alexander Sarris and Jamie Morrisson. Rome: FAO.

Ariga, J., T. S. Jayne, and S. Njukia. 2010. "Staple Food Prices in Kenya." COMESA Policy Seminar, Variation in Staple Food Prices: causes, consequence and policy options, Maputo, Mozambique.

Berman, Bruce J., Jill Cottrell, and Yash Ghai. 2009. "Patrons, Clients, and Constitutions: Ethnic Politics and Political Reform in Kenya." *Canadian Journal of African Studies/Revue canadienne des études africaines* 43 (3):462–506. doi:10.1080/00083968.2010.9707567.

Bohstedt, John. 2016. "Food Riots and the Politics of Provisions From Early Modern Europe and China to the Food Crisis of 2008." *The Journal of Peasant Studies* 43 (5):1035–1067. doi:10.1080/03066150.2016.1170009.

Boone, Catherine. 2014. *Property and Political Order in Africa: Land Rights and the Structure of Politics*. Cambridge: Cambridge University Press.

Cheeseman, Nic, Gabrielle Lynch, and Justin Willis. 2016. "Decentralisation in Kenya: The Governance of Governors." *Journal of Modern African Studies* 54 (1):1–35. doi:10.1017/S0022278X1500097X.

D'Arcy, Michelle, and Agnes Cornell. 2016. "Devolution and Corruption in Kenya: Everyone's Turn to Eat?" *African Affairs* 115 (459):246–273. doi: 10.1093/afraf/adw002.

Fengler, W., and J. Kiringai. 2009. "Still Standing: Kenya's Slow Recovery From a Quadruple Shock With a Special Focus on the Food Crisis – Kenya Economic Report." Kenya Economic Update. Washington, DC: World Bank.

Fengler, W., and J. Kiringai. 2010. "Running on One Engine: Kenya's Uneven Economic Performance With a Special Focus on the Port of Mombasa – Kenya Economic Report." Kenya Economic Update. Washington, DC: World Bank.

Gatimu, C. W. 2014. *Life in a Time of Food Price Volatility: Evidence From the Poor and Vulnerable in Mukuru and Lango Baya, Kenya.* Brighton and Oxford: Institute of Development Studies and Oxfam GB.

Höffler, H., and B. W. O. Ochieng. 2009. *High Commodity Prices – Who gets the Money? A Case Study on the Impact of High Food and Factor Prices on Kenyan Farmers.* Germany: Heinrich-Böll-Foundation.

Hossain, N., and D. Green. 2011. *Living on a Spike: How Is the 2011 Food Price Crisis Affecting Poor People?* Brighton and Oxford: Institute of Development Studies and Oxfam GB.

Jayne, T. S., and David Tschirley. 2010. "Food Price Spikes and Strategic Interactions Between the Public and Private Sectors: Market Failures or Governance Failures?" *Commodity Market Review* (2010): 122–138.

Jayne, T. S., Robert J. Myers, and J. Nyoro. 2008. "The Effects of NCPB Marketing Policies on Maize Market Prices in Kenya." *Agricultural Economics* 38:313–325.

Kanyinga, K. 2004. "Civil Society Formations in Kenya: A Growing Role in Development and Democracy." In *Civil Society in the Third Republic*, edited by D. Okello. Nairobi: National Council of NGOs.

Leys, C. 1975. *Underdevelopment in Kenya: The Political Economy of Neo-Colonialism* Berkeley and Los Angeles: University of California Press.

Lubaale, G. N. 2013. *Life in a Time of Food Price Volatility, Experiences of the Poor and Vulnerable in Mukuru and Lango Baya, Kenya.* Brighton: Institute of Development Studies.

Mamdani, Mahmood. 1989. "Social Movements and Constitutionalism in the African Context." CBR Working Paper No 2. Kampala: Centre for Basic Research.

Minot, Nicholas. 2011. "Transmission of World Food Price Changes to Markets in Sub-Saharan Africa." *IFPRI Discussion Papers 1059.* Washington: International Food Policy Research Institute.

Murphy, Sophia. 2009. *Strategic Grain Reserves in an Era of Volatility.* Minneapolis: Institute for Agriculture and Trade Policy.

Musembi, Celestine, and Patta Scott-Villiers. 2015. *Food Riots and Food Rights: the Moral and Political Economy of Accountability for Hunger in Kenya.* Brighton: Institute of Development Studies.

Ndegwa, Stephen N. 1996. The Two Faces of Civil Society: NGOs and politics in Africa. West Hartford CT: Kumarian Press.

Njagih, M. 2011. "Commodity Prices Spark Protests in Kenya." East African Standard, 20 April 2011. www.standardmedia.co.ke/business/article/2000033598/commodity-prices-spark-protests-in-kenya

Nyamu-Musembi, Celestine, and Samuel Musyoki. 2004. "Kenyan Civil Society Perspectives on Rights, Rights-Based Approaches to Development and Participation," *IDS Working Paper 236.* Brighton: Institute of Development Studies.

Okello, D. 2004. "Civil Society in the Third Republic: Synthesis of the Main Issues Emerging." NGO Council Conference on "Civil Society in the Third Republic" 15–16 December 2003, Panafric Hotel-Nairobi.

Otieno, David Calleb, Nathaniel Kabala, Patta Scott-Villiers, Gacheke Gachihi, and Diana Muthoni Ndung'u. 2016. "Men and Women of Words: How Words Divide and Connect the Bunge la Mwananchi movement in Kenya." *IDS Bulletin* 47 (1):99–112.

Republic of Kenya. n.d. (obtained July 2013). "Food Security District Profile: Kitui District, Eastern Province." Nairobi: Ministry of Agriculture, Government of the Republic of Kenya.

Republic of Kenya. 2007. *National Food Security and Nutrition Policy*. Nairobi: Government of the Republic of Kenya.

Republic of Kenya. 2010. *The Constitution of Kenya*. Nairobi: National Council for Law Reporting.

Republic of Kenya. 2011a. *National Food and Nutrition Security Policy*. Nairobi: Agricultural Sector Coordination Unit, Government of the Republic of Kenya.

Republic of Kenya. 2011b. *National Social Protection Policy*. Nairobi: Ministry of Gender Children and Social Development, Government of the Republic of Kenya.

Republic of Kenya. 2012. *Social Protection Sector Review June 2012*. Nairobi: Ministry of State for Planning, National Development and Vision, Government of the Republic of Kenya.

Scott-Villiers, Patta, Nick Chisholm, Alexandra Wanjiku Kelbert, and Naomi Hossain. 2016. *Precarious Lives: Food, Work and Care after the Global Food Crisis*. Institute of Development Studies. https://opendocs.ids.ac.uk/opendocs/handle/123456789/12190

Short, C., W. Mulinge, and M. Witwer. 2013. "Analysis of Incentives and Disincentives for Maize in Kenya." *MAFAP Technical Notes*. Rome: FAO.

Sinha, Dipa, Biraj Patnaik, Vaibhav Raaj, Shreya Bhattacharya, and Anuradha Joshi. 2014. *Popular Actions, State Reactions: The Moral and Political Economy of Food in India*. Brighton: Institute of Development Studies.

Thompson, E. P. 1971. "The Moral Economy of the English Crowd in the Eighteenth Century." *Past and Present* 50:76–136.

Tschirley, David L., and T. S. Jayne. 2010. "Exploring the Logic Behind Southern Africa's Food Crises." *World Development* 38 (1):76–87. doi: http://dx.doi.org/10.1016/j.worlddev.2009.09.008.

World Food Programme. 2011. *Kenya: An Evaluation of WFP's Portfolio 2006–2010*. Rome: Office of Evaluation, World Food Programme.

8 Authoritarian responsiveness and the *greve* in Mozambique

Luís de Brito, Egídio Chaimite and Alex Shankland

Introduction

In February 2008 and September 2010, Mozambique's capital Maputo and its satellite city Matola were the scene of violent protests by groups of ordinary citizens against rises in the cost of living, including the prices of bread and urban transport. Immediately afterwards, these protests were replicated in some other Mozambican cities, but on a much smaller scale, and they were quickly brought under control by the police. Although the media and the government described the protests using terms that ranged from 'uprisings' to 'vandalism', the population generally referred to them as *greves*, the Portuguese word for 'strikes'. In November 2012, a new *greve* was set in motion, but immediate repressive action by the police (which had been greatly strengthened by that time) in key gathering places prevented the protests from spreading, although this did not prevent a near total paralysis of activity in Greater Maputo for a day and a half.

Since then, Maputo has lived in uneasy expectation of the next *greve*. Expectation intensified in 2015, with spiralling debt crisis associated with the freezing or withdrawal of much of the international aid on which the country's ability to provide public services has historically depended. This triggered a devaluation of Mozambique's currency and consequent rapid rise in local food prices in import-dependent urban markets, aggravating the impact of a prolonged drought that had reduced local food production in 2015–2016. The government's response to the risk of another *greve* was a mix of repression and concession, an approach that staved off further protests in 2015–2016 despite deteriorating economic, social and political conditions.

In this chapter, we argue that this dynamic derives from the experience of the 2008 and 2010 *greves*, which has given the politics of provisions in Mozambique a distinctive character of 'authoritarian responsiveness'. This concept has emerged from studies of middle-income countries in Latin America, Eastern Europe and Asia – especially China, where there is a growing literature on the Communist Party's ability to combine repression with rapid course adjustments in the face of offline and online protests (Froissart 2014; Chen, Pan and Xu 2016). It has not yet been widely applied in studies of African politics, but we argue that it is a useful way of thinking about the distinctive ways in which Mozambique's ruling elite has responded to popular protests in a context where "processes, spaces and domains of the state are regularly challenged by formations beyond its control" (Bertelsen 2016: 2). In this chapter, we seek to show that, while the *greves* were part of a broader movement that has shaken many countries since 2008 as a result of global food price volatility, their characteristics and impacts have been shaped above all by local economic and political dynamics – and in particular by the interaction between the country's authoritarian political tradition and the social and political exclusion to which the poorest urban strata in Mozambique have been condemned.

This chapter is the result of an extensive process of primary and secondary data collection carried out between 2013 and 2014, with further analysis and follow-up work in 2015 and 2016. Beginning with a review of existing literature and secondary data on food security, urban poverty, wages and food price inflation in Mozambique, the study included the construction of a 'political event catalogue' based on newspaper reports of protests taking place between 2007 and 2012 and of the associated policy responses, and a process of discourse analysis which focused on comparing the coverage of the protests in government-controlled and more independent newspapers (Pereira *et al.* 2013). Primary qualitative data collection explored accounts of protest events in four popular neighbourhoods, two urban sites in Maputo and one rural and one urban site in Chókwe, a market town in the flood-prone Limpopo valley. This process included observation, individual interviews and focus groups, with the latter using participatory visualisation tools such as timelines and matrix scoring in addition to semi-structured group discussion.[1]

The chapter begins by presenting a brief historical perspective and a characterisation of the nature of the authoritarian political tradition that has marked Mozambique since the proclamation of independence in 1975, remaining dominant even while co-existing with a profound change in the political alliances of the ruling party. It then looks at the question of food and the cost of living, as well as the immediate context of the *greves*, the discourses and images used to describe them and the type of government responses that they received, reflecting on the extent to which these protests have revealed the contours of a contemporary Mozambican 'moral economy' – a term defined by Hossain and Kalita as including "(1) conceptions of material justice, (2) the idea of clear limits to the tolerance of exploitation and unfairness, and (3) the impacts of sweeping economic and political change on understandings of fairness and equity" (2014: 4). We go on to examine how Mozambique's "politics of provisions" – Bohstedt's term for "the ways in which common people interacted with their rulers over subsistence" (2016: 1036) – has been shaped by the failure of civil and political society in Mozambique to institutionalise the popular demands that found expression in the *greves* via the formal spaces for debate and negotiation of the country's political system. We conclude by analysing the extent to which the protests have produced new accountability relations – however informal and unstable – between Mozambique's government and its citizens, as well as the role in shaping these relations that is played by the country's dependence on international donors.

Authoritarian tradition and political alliances

The recent political history of Mozambique is characterised by an authoritarian political tradition, both under the one-party state, which lasted from independence in 1975 until 1990, and in the period that followed the introduction of a multi-party system and the end of the civil war. Indeed, after a brief period of weakening and hesitation in the years following the General Peace Agreement in 1992 and the first multi-party elections in 1994, the ruling Mozambique Liberation Front (FRELIMO, or Frelimo) chose, from 2002, to strengthen the party-state model, which had developed during the one-party period. To this end, the network of party branches and committees was reestablished throughout the state apparatus (including the defence and security forces), from the central to the local level. The principle of accountability of state officials to Frelimo party leaders at the various levels was also reactivated.

In the period immediately after independence, despite the fact that Frelimo had fought its guerrilla war among the peasantry and with a discourse evoking the 'worker-peasant alliance', its political choices resulted in a de facto alliance with urban strata, particularly middle- and low-income groups. Indeed, even in a context of crisis, it was these groups who most benefited from Frelimo's price control policies, as well as from its measures to promote access to services, especially health, education and housing. The peasant population was severely affected by the breakdown of the agricultural marketing system and by the civil war, which substantially disrupted food production conditions and led to millions of Mozambicans becoming internally displaced people or refugees.

Under pressure from the international financial institutions, Frelimo adopted a policy of economic liberalisation in the mid-1980s, but due to the war this had practically no significant effects. From the 1990s, however, it was resumed with price liberalisation and the privatisation of most state-owned companies. This period saw the start of a search for foreign investment, accompanied by the offer of fiscal incentives, which would lead a few years later to the emergence of a series of mega-projects, mostly linked to exploitation of natural resources (principally natural gas, coal, heavy mineral sands and oil). Alongside economic liberalisation, the Frelimo leadership underwent a conversion to capitalism in its own affairs and became increasingly involved in the world of business. Thus, the party's old political alliance was replaced by a new one, this time with business circles, and particularly with large-scale foreign capital.

Throughout this period, the authoritarian tradition that characterises Mozambican politics and the hegemony of Frelimo (which has always enjoyed an absolute majority in parliament) have been major obstacles to the development of alternative political projects and to the scope for citizens to organise themselves and freely express their opinions, interests and demands, either individually or collectively. Also contributing to this is the fact that the government – and thus Frelimo – exercises direct control over the main publicly-owned mass media.[2] Additionally, the government is able to influence many of the more recently-established media outlets (newspapers, television and radio) through its power to control the advertising flows on which these outlets depend.[3] This has inevitably limited the space for the appearance of an informed and critical public opinion and for open debate on questions such as the rise in the cost of living in general and in food prices in particular, despite the status of these issues as pressing concerns for ordinary citizens.

By the time of the 2010 *greve*, the picture of Frelimo media hegemony had begun to change a little with the growth (principally in urban areas) of a range of new informal spaces for political debate resulting from increasing access to social media (particularly Facebook) and the emergence of a new generation of young people with relatively high levels of education (including a growing number with university degrees). This generation provides a potential constituency for Mozambique's abundance of civil society organisations (CSOs), but (as we discuss later in this chapter) few Mozambican CSOs are fully independent of Frelimo, and almost all remain heavily dependent on funding from foreign donors.

The potential voting power of educated urban young people became clear over the course of a series of local elections between 2009 and 2013 in which a relatively new opposition party, the Mozambique Democratic Movement (MDM), won control of some of Mozambique's largest cities by capturing a large share of the youth vote. However, while formally tolerating opposition parties and the structures of multi-party democracy, Frelimo regards the former as politically illegitimate. This attitude has shaped both Frelimo's efforts to restrict the freedom of operation of MDM mayors in the cities and its reluctance to embrace a negotiated peace agreement with the largely rural-based armed opposition group Renamo (Mozambique National Resistance). Since 2013 Renamo has combined an intermittent insurgency in the centre and north of the country with continuing to operate as a political party, and performed strongly in the 2014 presidential election.

The way in which Frelimo's control has made it almost impossible to develop structures of collective representation that might be considered by the regime as legitimate interlocutors is also evident in the fact that it has hollowed out Mozambique's constitutional guarantees of the rights to freedom of assembly and of demonstration. The government requires those organising demonstrations to seek prior authorisation, and while marches directly or indirectly promoted by the ruling party or organisations close to it tend to be permitted, in most cases those led by independent groups or opposition parties are denied authorisation and/or repressed by the police in a particularly violent way. The lack of space for the free exercise of democratic rights has resulted in severe limitations on the room for social negotiation and on citizens' experience of and access to different forms of mass political mobilisation (such as petitions, boycotts, strikes or marches). This leaves the low income urban population, as it seeks to respond to the effects of sudden increases in the cost of living, with few alternatives but to resort to violent street protests that take place without organisational leadership. In later sections of this chapter, we examine the extent to which *greves*, such as those which took place in 2008 and 2010, are part of an emerging 'repertoire of collective action' in Mozambique (Chaimite 2014). First, though, we examine the contextual drivers of food price volatility in Mozambique, and the ways in which food insecurity has been experienced by the citizens of the popular neighbourhoods where the 'Food Riots and Rights to Food' study was conducted.

The food question and the cost of living

Although Mozambique has always been a food importing country, the situation was particularly serious during the civil war period (1976–1992). In 1975, the year of independence, 49 per cent of the grain supply was produced domestically, 42 per cent came from commercial imports and 9 per cent came from food aid; but by the end of the 1980s, domestic production accounted for only 14 per cent, with the remaining 86 per cent coming from food aid (Eduardo Mondlane University Economics Faculty 1990: 13). Food aid continues to play an important role in supplying the country, as although domestic food production has risen since the end of the war, the country remains far from self-sufficient. In particular, Mozambique has a situation of persistent deficit in the production of rice and wheat, imports of which account for the majority of the annual general grain deficit of around 17 per cent (SETSAN 2013).

Food production in Mozambique, the vast majority of which comes from peasant family farming, has long been affected by the natural disasters (including droughts, cyclones, floods and pest outbreaks) that frequently strike the country, triggering government emergency programmes backed by foreign aid. Government policy documents have long placed great emphasis on agricultural production and productivity, but their impact has been practically nil, as shown by the feeble growth of the agricultural food sector in recent years and the stubborn persistence of high poverty levels in the country.[4] According to Mosca (2014), per capita domestic food production has been declining in absolute terms over the last fifty years, and for a significant number of the main food crops (with the exception of maize) productivity per hectare has also been declining.

Contrary to expectations that national food production would increase in response to the rise in prices for food on the international market in the 2007–2013 period, this rise had practically no impact on the domestic production of foodstuffs. According to producers we interviewed in Chókwe, production during this period was strongly affected by increases in the cost of inputs, by climatic conditions (floods and drought), as well as by uncertainty over the extent to which a market for their crops could be guaranteed, undermining the stimulus effect of rising food prices.[5]

Meanwhile, the overall food security situation deteriorated significantly. Most households in our research sites reported scaling back their food consumption to one main meal at the end of the day, with a few leftovers being carefully saved for breakfast. The major difference we found between rural and urban areas is that people in the latter relied overwhelmingly on bread, made from imported wheat flour (obtained both on the international market and via food aid) supplied to bakeries whose prices were under tight government control. Bread was usually eaten on its own or as a sandwich with a leaf of cabbage or slice of tomato – eggs, fish or chicken were considered rare and increasingly unaffordable luxuries. The traditional rural diet, based on maize and cassava, had been rejected by urban Mozambicans for both aesthetic and practical reasons: it was seen as rustic and unsophisticated, and also required expensive charcoal and scarce time to cook, whereas bread could be eaten as soon as it is bought.

The food security situation of the poor in the peripheries of urban areas, where the riots of 2008 and 2010 started and were most intense, is particularly difficult. That is because unlike rural areas, where household production accounts for about 75 per cent of the food consumed, in the peripheries of Maputo and Matola the poorest families rely on cash income flows for 97 per cent and 91 per cent of their food needs, respectively (Mandlate 2015). The vulnerability of poor households in these two cities to fluctuations in food prices is not mitigated by government subsidies, which according to Mandlate (*op. cit.*) principally benefit wealthier households.

When asked to identify who was responsible for the degradation of their food security situation, people in these areas were outspoken in their contempt for the ruling elite. However, they sharply differentiated the recent period of Frelimo rule from its previous incarnations – especially the heroic socialist period under Samora Machel, in which hunger was remembered as an experience that was shared by rulers and ruled, as government ration shops distributed the meagre food supplies that were available in an economy ravaged by civil war and agricultural collapse. The nostalgic refrain 'in Samora's day . . .' was a recurrent one in our interviews and focus group discussions, used to refer to a period that not only symbolised greater economic equality but also official intolerance of corruption and self-interest among government officials. It was clear from our interviews that Mozambique's dramatic recent rise in income inequality, which doubled in the six years to 2015 (Hanlon 2016), was associated with a perceived increase in corruption among the ruling elite.

Another refrain was the popular phrase *quem não trambuca não manduca* – 'those who do not work, do not eat'. The moral economy in both rural and urban Mozambique is strongly marked by the understanding that food security is something to be earned through hard work by everyone except the very elderly or infirm; an understanding that resonates with the deep-rooted experiences of forced labour regimes under pre-colonial conquerors, Portuguese colonists and post-colonial state and rebel forces alike that are powerfully analysed by Bertelsen (2016). Frelimo's liberation promise that the state would deliver material progress in return for hard work continues to hold broad popular legitimacy; within this logic, in urban Mozambique the government's violation of the moral economy is perceived as lying not in a failure to distribute free food but in a failure to preserve the relationship between wages and food prices. The result of this failure is that the idle children of the elite grow fat while even the hardest-working semi-skilled labourer cannot earn enough to feed a family – a situation described by one middle-aged man in the Maputo neighbourhood of Ferroviário as a "psychological torture" perpetrated by the government.[6]

Focus group participants recalled that food prices had also risen in the 1990s under another Frelimo president, Joaquim Chissano, but insisted that during that period wages had risen enough to keep pace, whereas since 2008 they have been outstripped by inflation. Indeed, as shown by Wuyts (2011), the period since 2007 has seen the annual average rate of food price inflation clearly outstrip the rise in prices for other products. Inflation has thus had a much greater impact on the urban poor than on wealthier groups who tend to spend a smaller proportion of their income on food, further intensifying the effects of inequality in Mozambique (Arndt, Jones and Salvucci 2015). However, governments in power during this period have consistently refused to recognise this, and labour union representatives interviewed during our study accused the government of short-changing workers by insisting that any wage rises should be linked to a complex measure of sectoral GDP growth, rather than to the cost of a 'basic basket' of goods in which food items would be heavily represented.

Popular revolts

Since 2007, Mozambique has experienced several explosive episodes of popular mobilisation (in 2008, 2010 and 2012) in which protesters have voiced anger not only at the rising cost of urban transport and food but also at growing inequality and ever more visible corruption. Although these revolts spread to other towns and cities (including Chókwe, where we interviewed people who had protested in the local marketplace), their birthplace and main stage was Greater Maputo, a metropolitan area covering the country's capital and its satellite city of Matola, with a combined population of about two million people. Greater Maputo is highly unequal; its average indicators of development and wellbeing, higher than those for any other region of Mozambique, disguise the fact that it also contains the country's largest concentration of people living in poverty. Maputo is also highly connected, with levels of education and rates of computer, internet or cell phone access that are well above those for the rest of Mozambique. As noted by Tvedten *et al.*, the perception of "being trapped in poverty and of being unable to take part in what the city has to offer, is at the heart of the current dissatisfaction among poor people in the poor neighbourhoods of Maputo" (2013: 4). Greater Maputo is also a region where support for the ruling party has historically been very strong, with Frelimo consistently obtaining around 90 per cent of the vote in Maputo and Matola between the first multi-party elections in 1994 and the 2004 general election (the last to take place before the 2008 protests).

In early 2008, the Mozambican government decided to increase the price of liquid fuels to compensate for the increased cost of imported oil (World Bank 2008). This decision led the operators of private passenger transport vehicles (known as *chapa 100* or simply *chapa*) to increase their fares by between 33 per cent and 50 per cent. In this same period, there had already been an increase in the price of some foodstuffs, notably bread, the price of which had gone up by 12.5 per cent. After several days in which text messages circulated widely calling for a *greve* or "strike", on 5 February a wave of violent popular protests erupted, which included blocking the main roads with improvised barricades and burning tyres, looting shops (particularly of food), and burning or stoning vehicles in the places with the largest concentrations of protestors. The riots completely paralysed the cities of Maputo and Matola for three days.[7] The protests ended when the decision was taken to keep transport fares unchanged, which was possible after the government negotiated a series of compensatory benefits with the *chapa* operators, notably the removal of the 17 per cent value added tax on diesel.[8]

On 1–2 September 2010, after text messages again circulated calling for a *greve*, a second major outbreak of street protests occurred. This time the protests came in response to the announcement of an increase in the government-controlled price of bread and other basic goods and services.[9] The prices of foodstuffs such as rice, tomatoes and onions had already seen significant rises, as had that of fuel. Similar patterns of violence to those observed in February 2008 were seen, this time with around ten deaths and hundreds of people injured, in addition to extensive material damage. And, just as in 2008, the government ended up reversing its decisions, freezing prices and announcing subsidies, among other measures.

On 15–16 November 2012, Maputo was paralysed again, this time for a day and a half due to the threat of riots following an announced increase in transport fares. However, the patterns of violence seen in 2008 and 2010 did not recur. Mobile phone communications were restricted, and a significantly strengthened police force maintained a heavy presence at all the critical points of the city that had acted as foci for previous protests.

These foci included neighbourhoods such as Inhagoia, Chamanculo, Benfica, Polana Caniço, Maxaquene and Magoanine, in which large numbers of people who feel trapped and cornered by poverty are found. In particular, they have large concentrations of young people in *waithood*, described by Honwana as a kind of "prolonged adolescence or an involuntary delay in attaining the adult state" (2013: 5). Waithood is experienced by youth who have precarious jobs or no formal employment and are unable to build, buy or rent a house, or to marry, form and feed a family. In the neighbourhoods where we conducted the fieldwork for this study, it was above all young people in waithood who had been involved in the protests.

The protests always had an anonymous origin. Mobilisation was achieved through messages which circulated virally across the mobile phone networks and on social media (in particular, Facebook), announcing the *greve* or 'strike'. The use of the term *greve*, the word most frequently used by participants and by ordinary citizens to describe the protests, can be understood as identifying the movement as one which made claims upon a 'boss' – in this case the government, which was held responsible for guaranteeing all citizens the minimum conditions for a decent life.

As one interviewee in the neighbourhood of Chamanculo put it:

> It so happens I received it by a text message. But if you ask me where the message came from, I won't know how to answer. I just received the message.

> That is to say there was a means of spreading people's spirit of dissatisfaction via messages. We also have the social networks. You have to know that these means are great platforms for spreading information. And the people, in these social networks, give their opinions about the cost of living . . . the spirit of the messages threw stones at the current government . . . they said it was a government of intransigent, intolerant individuals who had nothing to do with the cause of the people or with democracy, because one of the pillars of democracy is full respect for human rights. The right to health and the right to food are part of human rights.[10]

The authorities considered that the popularity of protest music had also been a factor in triggering the demonstrations. The public prosecutor's office accused one musician, Azagaia, of inciting the 2010 *greve* with his song *Povo no Poder* ("Power to the People"), which deals with the difficult situation experienced by the Mozambican people faced with the increase in the cost of living. The lyrics of this song, recorded after the *greve* of 2008, include the following verse:

> Mr President, you left the luxury of your palace
> You finally noticed that life's not easy here
> Only now did you call a meeting of your Council of Ministers
> But the people haven't been sleeping, we came together a long time ago
> We've barricaded the streets
> We've halted the minibuses
> No one is getting past
> Even the shops are shut
> If the police are violent
> We'll respond with violence[11]

Despite the visibility of young people at the forefront of the protests, they were by no means the only ones to take part. Our interviewees said that, directly or indirectly, a sizeable number of residents in the places that were the foci of the protests joined in, and that many households benefited from the foodstuffs looted from shops and warehouses, even including the (notoriously underpaid) police themselves. According to an interviewee from the neighbourhood of Ferroviário:

> There are households here who went for months without buying rice, because they stole a lot of sacks. And some households sold the stolen rice at a cheaper price . . . This just showed that the price had to drop . . . The police are also human. They see that the people are right. It's the situation . . . This is the reality. Some policemen also stole, because they needed to.[12]

Among our interviewees people's participation in the protests was justified on the grounds that they were hungry, and blame for the *greves* was attributed largely to the government which did nothing to improve the situation; for example, by creating jobs for young people and ensuring that workers' wages could keep up with the cost of living. However, in interviewees' discourses, joining the protests (and the looting) was presented a momentary solution to the problem of lack of food, rather than as a part of a more strategic or sustained attempt to address the causes of the problem. Despite the perceived success of the *greves* in securing temporary relief from worsening food insecurity, the lack of sustained engagement suggests that Mozambique has not embarked on the kind of historical pattern described by Bohstedt, under which "where the politics of provisions brought relief repeatedly . . . it contributed to regime stability by incorporating the interests of popular actors with leverage into the political process" (2016: 24). Although, as Bertelsen notes, the *greves* "generated novel collectivities" (2014: 21), they were above all ephemeral, unorganised phenomena, which did not result in any more durable form of organisation.

Nonetheless, in addition to their dimension of raising immediate claims, these revolts can be seen as a form of political expression, bringing new elements to the repertoire of political action in Mozambique (Chaimite 2014)[13] – a repertoire that, as it becomes broader, shapes the strategies of both the potential participants and mobilisers and the political-administrative authorities. The repertoires of other segements of society have also been evolving: for example, after the violent demonstrations of 2008 and 2010, and the blocked attempt to stage another *greve* in 2012, a largely middle-class civil society protest against a crime wave and the threat to peace resulting from Frelimo's escalating confrontation with Renamo in 2013 saw the first major peaceful demonstrations against the government in Mozambique's recent history. In the next section, we examine how the government's own repertoire of responses has evolved since 2008.

From confrontation to authoritarian responsiveness

The authorities were apparently taken by surprise by the protests of 2008, and the police response was slow and ineffective. In 2010, the government moved more quickly when people began to flood onto the streets in response to text messages calling for a *greve*, but the indiscriminate violence of the police response served mainly to incense the protesters further, as the firing of live rounds into the crowd led to at least fourteen deaths and hundreds of injuries.

On the eve of the 2010 protests, the Ministry of the Interior had tried to demobilise people by saying that the strike would not take place because there had been no notification of a demonstration as required by law. The President of the Republic called for vigilance against "agitators and the preparation or undertaking of acts which constitute an assault against life or against order and public tranquility."[14] As the protests began to spread, the Minister of the Interior described them as "illegal", "violent" and instigated by "adventurers, unemployed people and bandits" who were hiding behind "innocent citizens, mostly children."[15]

The government ordered mobile phone network operators to suspend the pre-paid text message service and a few days later introduced a compulsory registration programme, "to promote the responsible use of SIM cards, thus contributing to maintain public order and tranquillity."[16] Initially registration was required within a period of two months, after which unregistered numbers would be blocked.[17] The measure was criticised by civil society groups as an attempt at "repression of the freedom of expression which potentially contests social inequalities" (CIP 2010). Although SIM card registration was common practice in many countries, the timing and the context of the decision to apply it in Mozambique emphasised the potentially repressive character of the measure. This is borne out by the fact that although the government initially backed away from enforcing this measure as the memory of the 2010 *greve* receded, the registration deadline was reintroduced in 2012, when there were new threats of protests after the rise in public transport fares in Maputo.

During the *greve* of 2010, the Minister of the Interior insisted that the authorities would not yield to pressure to reverse price hikes that he described as resulting inevitably from a global rise in the price of oil. But as the combination of aggressive rhetoric and violent police repression failed to demobilise the protesters, coverage in the non-state media, social media discussion and even statements by Frelimo's own local representatives began to signal widespread understanding of the reasons for the protests, given the evidence of the impacts of the rising cost of living on the poorest strata. Faced with the strength of the street protests and a rising tide of disapproval among the non-poor, the government switched – as it had in 2008 – from a strategy of confrontation to accommodation through concessions and new policy announcements.

Since the trigger for the 2008 and 2010 riots had been a sudden hike in *chapa* fares, in the first instance the government froze the price rises and promised to compensate the transporters, provided they showed their operating licence, regularised their tax situation and registered their fuel consumption. The 2010 concessions were more comprehensive. On 2 September 2010, at the end of the second day of violent rioting, President Guebuza not only announced immediate subsidies to transporters and bakers, but also a freeze on the wages and allowances of officials of state agencies and public companies.[18] He explained that "to contain the impact of these crises on the lives of citizens" and "stepping up its actions in the fight against poverty in the cities and the countryside", the government would be implementing a food production action plan, introducing a "basic basket" of subsidised goods, increasing urban District Development Funds,[19] reducing customs duties on the import of third grade rice and of sugar, and promoting the production and consumption of bread made from cassava flour. This became the 'Package of Measures to Mitigate the Cost of Living' approved by the Council of Ministers on 29 March 2011 (Governo de Moçambique 2011).

The Passenger Transport Subsidy Programme element of the package was a strategy "to liberalise gradually the transport fares paid by passengers, but seeking to protect the most vulnerable strata of the population, and/or the productive functions which could suffer from an increase in fares" *ibid*. The programme emphasised both the government's alignment with criticisms of subsidies by the international financial institutions and its paternal concern for the welfare of needy citizens, noting that the old subsidy model:

> is not directed at those who need it most . . . it covers everyone, from the student or worker going from home to work and back to the passenger going to

> the beach or going shopping. This distorts the economy and could destabilise the macro-economic environment which determines growth and harmonious development of the country in all its dimensions, . . . even putting at risk justice in the distribution of income.
>
> (*ibid.* p. 21)

Although the new subsidy system was to be introduced before December 2011, it was never implemented. Transport continued to be subsidised through payments to *chapa* fleet operators, but as the years went by, transporters were also required to pay increasing licence fees and taxes. The deputy chair of the Maputo Association of Road Transporters explained that this demand for formalisation meant that transporters were claiming the subsidy less and less: 'it becomes a process whereby you take money out of one pocket and put it in another."[20] As a result, rising fuel prices once more led to *chapa* price increases in 2012, triggering calls for another *greve*; although (as noted previously) it was suffocated by a powerful and rapidly-mobilised police presence, the threat of a new *greve* nonetheless paralysed Greater Maputo for a day and a half as people stayed at home to avoid being caught up in confrontations.

Another measure promised by the March 2011 package was the subsidised 'basic basket'.[21] Those who were to benefit were initially defined as formal sector workers living in the eleven provincial capitals who earned a wage equal to or lower than 2,000 meticais per month (then equivalent to roughly USD 65). Shortly afterwards, this was extended to include all informal sector workers and all those who received a monthly wage no larger than 2,500 meticais (USD 80). The measure envisaged that the beneficiaries could buy, at subsidised prices, a series of basic goods – namely rice, maize flour, vegetable oil, second grade fish, butter beans, sugar and bread.

This measure was criticised as inconsistent and unsustainable, since the exact number of beneficiaries was unknown and no feasibility study had been undertaken; it was also attacked for being discriminatory since it would only benefit the inhabitants of urban areas, which had been the epicentres of the protests. In the end, the measure was never implemented; the government claimed in the period immediately after legislating for the 'basic basket' in 2011 that one of the preconditions for its implementation – the occurrence of further increases in grain and fuel prices – had not been met. In fact, grain and transport prices did rise in Mozambique after 2011, despite the relative stabilisation of fuel prices on the international market. However, until recently these rises were relatively gradual, reflecting the government's desire to avoid large and sudden price rises with their attendant potential for conflict. Even the acceleration of inflation in 2015–2016 failed to reactivate debate on the 'basic basket' or government commitment to its application, suggesting that it has been comprehensively discredited as a policy response.

Government announcements of measures to improve food production, storage and use proved equally ineffective. In 2008, a strategic food storage programme based on grain silos located in key agricultural regions was announced, but by

2012 only two of the promised 39 silos had been built. At the time, domestic production of wheat in Mozambique was insignificant, but consumption was rising, particularly among the poorest strata of society in urban areas, with the result that variations in the price of bread had a significant impact on their lives. In 2008, a Food Production Action Plan (PAPA)[22] was introduced to reduce the country's vulnerability to price variations on the international market. According to this plan, national wheat production should have increased fivefold by the 2010–2011 agricultural season, reducing the wheat deficit from 95 per cent to 20 per cent of total consumption. It was also hoped that cassava production would increase fivefold over the same period, while a new technique was to be developed to bake bread from a mixture of wheat and cassava flour.[23] These efforts notwithstanding, the production of wheat and other grains remained virtually unchanged over the following five years. The deficit was made up through food imports, mainly from South Africa, and, in the case of wheat and rice, through food aid donations from the USA and Japan.[24]

In 2010, the national currency, the *metical*, fell by some 14 per cent against the dollar, with very strong effects on the price of fuel and grain on the domestic market. The Bank of Mozambique moved to guarantee the foreign exchange needed to finance fuel and food imports at a more favourable exchange rate, and succeeded in shifting the metical/dollar exchange rate from 37 in late August 2010 to 31 in March 2011.[25] Tight control of the exchange rate came to operate from then on as an important instrument to keep food prices down, and remained effective until the debt crisis took hold in 2015–2016. While the metical/dollar rate was still around 34 at the start of 2015, by the beginning of 2017 it had risen above 70, signalling a fresh cycle of intense pressure on food and transport prices.

Considering both repression and accommodation, this section has shown the form taken by authoritarian responsiveness in Mozambique between 2008 and 2012. The politics of provisions in this period saw a series of often-violent protests, repressed with ever greater force, balanced by what appeared to be increasing sensitivity to the concerns of the urban poor. This sensitivity was reflected both in policy announcements that were largely discursive performances with little or no tangible impact (classified as *mentiras* or 'lies' by many of our community-level interviewees) and in concrete measures to attempt to maintain price subsidies and a favourable exchange rate, even in the face of intense macroeconomic pressures. In the next section, we examine the failure of other actors involved in Mozambique's political and policy debates to press government to turn ad hoc responses to the *greves* into to a more institutionalised approach to preventing hunger and tackling gross income inequality in Mozambique.

Responses from political society

Bohstedt argues that a country's "politics of provisions" is shaped by forms of bargaining between rulers and ruled that are marked by displays of force but may also be mediated by "existing networks – of solidarity among the common people and reciprocity between them and their rulers" (2016: 1). His examples of reciprocity include basic welfare provision by elites in 18th century England and China, as well as the 'bread bargain' that shaped political relations in Arab countries between the 1970s and the uprisings of 2011. As discussed in earlier sections of this chapter, in Mozambique there is a popular imaginary of reciprocity between rulers and ruled associated with memory of the post-independence socialist period; however, the expectation of reciprocity between government and citizenry has long since been undermined by rising inequality and ever-more-visible corruption among the political elite.

Although all three main political parties – and in particular Frelimo – may sometimes make formal mention of agriculture as a priority area in their election manifestos, this is a purely rhetorical commitment rather than a genuine expression of social interests.[26] As discussed previously, Mozambique's political system as a whole is marked by the authoritarian political tradition that was inherited from the colonial period and further developed by Frelimo during the period of one-party rule. Frelimo's historically vanguardist, elitist and top-down characteristics are also evident in the political cultures of Mozambique's other political parties (most of which were formed by breakaway groups of former Frelimo members). As a result, party structures function basically as clientelistic networks and as transmission mechanisms for 'guidance' issued from the higher echelons to the lower, leading to a virtually nonexistent capacity to respond to citizens' concerns and represent them effectively.

This situation is exacerbated by Mozambique's parliamentary system, based on a form of proportional representation with closed party lists. This system displaces the focus of politicians' energies away from developing relationships with voters (who are only called upon to vote for their preferred party once every five years, and are generally unaware of which candidates are actually on a given party's list) towards developing relationships within the party, since it is party colleagues – and especially leaders – who will determine whether their name is placed close enough to the top of the party's list of candidates for them to have a chance of winning a seat. The result of this combination of an authoritarian political tradition with a system of political representation controlled by the party machines is a drastic reduction in the scope for effective democratic representation.

The convergence of factors has resulted in a situation where political parties, far from providing the citizens of precarious urban neighbourhoods with an alternative to the *greve*, are a virtual irrelevance in Mozambique's politics of provisions. Although we do not go into it here, CSOs have also proven unable to represent and articulate demands for food rights, above all those of the urban poor. As Bertelsen puts it, "the notions of effervescence and ephemerality that characterised the popular uprisings are key aspects of a form of post-colonial politics . . . that is irreducible to the orderly world of representation (through NGOs, trade unions and political parties, for example)" (2014: 21). In Bohstedt's terms (2016: 1035), within Mozambique's politics of provisions, "solidarity among the common people" is expressed through powerful but ephemeral protests, and "reciprocity between them and their rulers" is sought through a bargaining process centred on force rather than democratic mediation.

Between responsiveness and repression

As we have seen, instead of the consolidation of a repertoire of peaceful and democratic modalities of citizen-state dialogue on poverty and food insecurity, the period since the *greves* of 2008 and 2010 has been marked by an unstable equilibrium between responsiveness and repression. This has persisted since the most recent episode of citizen-state confrontation around these issues, the abortive *greve* of 2012, enduring through 2016 and into 2017 despite the ongoing deterioration in the living standards of the poorest segments of the urban population, now intensified by the debt crisis and the resulting significant devaluation of the metical. The significant increase in levels of dissatisfaction among the urban poor has been met by a classic display of authoritarian responsiveness, as the government has combined measures intended to limit the social impact of the crisis with a significant investment in the state's repressive capacity.

Although nothing has come of the longer-term responses to the deteriorating food security conditions of the urban poor outlined in this chapter – including the government's promises to provide a subsidised 'basic basket' and to make food production a priority – there has been a concerted effort to hold down the prices of key commodities. This has included not only maintaining the subsidies on fuel for urban transport providers and wheat flour for bakeries, and blocking attempts to raise bread and transport prices, but also reducing import duties on basic foodstuffs in an attempt to offset the impact of the deteriorating exchange rate on food prices.

It is significant that although these policies directly contradict the standard policy recommendations of the World Bank and IMF, the latter – which has come to play an increasingly central role in attempts to tackle Mozambique's ongoing debt crisis – has not pushed hard for their reversal. Although documents from the international financial institutions continue to make reference to the need to eliminate subsidies, this is offset by a simultaneous emphasis on the need to protect the poorest in society. Together with this tolerance on the part of the financial institutions, the willingness of donor countries to continue providing food aid has helped to reduce the pressure on the government.

The maintenance of such provisioning-oriented measures has not, however, been the principal element of the regime's response to the *greves*. In fact, since 2008, Frelimo has acted in line with the authoritarian logic that has characterised most of its period in power and which has intensified over the course of the last decade, seeking to reinforce the repressive capacity of the state. This has centred on an accelerated process of development and equipping of paramilitary police forces (some of which have been used in the armed confrontation with Renamo in the central region of the country) and on reinforcing the security services and the Presidential Guard. It has also included efforts to exercise preventive control over any attempts to organise a demonstration, which have included installing video surveillance cameras at strategic points across Maputo and forcing mobile phone companies to block all unregistered SIM cards. The growing strength and effectiveness of the security apparatus suggests that Frelimo may be preparing for a shift away from authoritarian responsiveness – with its careful balancing of threats and concessions – towards a form of authoritarianism that relies on more effective repression to relieve it of the need to offer concessions at all.

The depth of the current financial crisis has pushed Mozambique back into a situation of heavy dependence on donor countries and the IMF, which is likely to last until the country's oil, gas and coal revenues have grown enough to cover the financial needs of the state. This seems to have resulted in a kind of tacit bargain to internationalise accountability for hunger in Mozambique, catalysed by the threat that the explosive combination of intensified popular anger and stronger state capacity for repressive violence means that any new *greve* risks resulting in even worse bloodshed than the previous ones. Under this bargain, donors agree to allow the government the political and fiscal space to maintain some measures to cushion the impact on the urban poor of food and fuel price rises, and Frelimo agrees to preserve a minimum of democratic space.

In her analysis of authoritarian responsiveness, Froissart argues that "authoritarian regimes last in part thanks to certain forms of discontent by showing that the way they are expressed is an integral part of authoritarian governance" (2014: 219). Mozambique's regime, however, seems set on stamping out the space for any expressions of discontent, whether they take place via the street protests of the *greve* or within what Bertelsen calls the "orderly world of representation" (*ibid.*). This reversion to Mozambique's cyclical pattern of periods of partial political opening followed by reassertion of an authoritarian governance model has been interrupted since 2016 by the debt crisis, with its accompanying internationalisation of the country's politics of provisions. There is little evidence of the emergence of non-violent channels for pursuing accountability for hunger at the national level in Mozambique, but the space preserved through this temporary arrangement may yet allow such channels to become established; we must recognise that the die is not yet cast.

Notes

1 Primary and secondary data collection for this study was led by IESE, the Institute of Social and Economic Studies (Luís de Brito, Egídio Chaimite, Crescêncio Pereira, Lúcio Posse and Michael Sambo) with the support of IDS, the Institute of Development Studies (Alex Shankland). Maputo data collection was also supported by Albino Xinavane, Elisa Machel and Halima Haji, three local young people who had been trained in audiovisual documentation by IDS researchers working with the DFID-funded *Diálogo* urban governance programme. The study was part of a four-country project entitled 'Food riots and food rights: the moral and political economy of accountability for hunger', led by Naomi Hossain (IDS) with colleagues in Bangladesh, India, Kenya and the UK and funded by the Economic and Social Research Council / Department for International Development Joint Fund for Poverty Alleviation Research (Grant reference ES/J018317/1). We are grateful to participants in the national reference group meetings – especially Edgar Cossa, Márcia Cossa and Daniel de Sousa – for their invaluable input into the analysis process. We would also like to thank the many people in Maputo and Chókwe who gave up their time to be

interviewed and to join focus group discussions. Not all of these people can be named at this time for a number of reasons. We would also like to thank our colleagues from Kenya, India, Bangladesh and the UK for the valuable feedback and stimulating discussions in the project workshops in January 2013 and February 2014 and in many subsequent Skype and email discussions.

2 This includes the national radio station (*Radio Moçambique*) and the public television channel (*Televisão de Moçambique*), as well as the longest-established newspapers (*Notícias* and *Domingo*); the latter belong to a company in which the main shareholder is the Bank of Mozambique.

3 Not only are the state and the public companies responsible for a significant part of the advertising, but Frelimo also has influence in other large companies, given the prominent positions that members of the Frelimo leadership occupy on their boards of directors, including of the major multinational companies working in the country.

4 According to van den Boom (2011), after a reduction in poverty in the years immediately following the end of the civil war (from 69 per cent to 54 per cent between 1997 and 2003), from 2004—2009, the levels of poverty remained practically unchanged (at 54 per cent).

5 The agribusiness firm MIA succeeded in boosting rice production in Chókwe significantly in 2008–2009, based on conditional provision of improved rice seeds to peasant farmers and a guarantee to purchase their output at a given price. There were also no floods during this growing season, which helped to guarantee a good harvest. However, there was heavy flooding the following season, leading to a fall in output from which rice production did not fully recover over the following seasons.

6 Interview, Ferroviário, August 2013.

7 Protest initiatives which broke out in some other towns and cities were immediately repressed and stifled by police action. Although there are no exact statistics, it is thought that at least 10 people died in the police action and that the cost of the damage was in excess of 100 million meticais (USD 4.2 million).

8 In principle, this measure applied only to operators who could prove they were licensed, had their tax situation regularised and recorded their fuel consumption. It is not clear how it was effectively applied to the large number of transporters whose situation was not fully regular. Apparently, the practice of 'route shortening' (interruption of the agreed journey before reaching the terminal), which has been growing since then, was the method the transporters found for compensating for the frozen fares.

9 See: http://www.telegraph.co.uk/news/worldnews/africaandindianocean/mozambique/7975093/Six-dead-in-Mozambique-riots-over-food.html;http://www.bbc.co.uk/news/world-africa-11150063; http://www.independent.co.uk/news/world/africa/seven-killed-as-rising-food-prices-spark-riots-in-mozambique-2068107.html

10 Interview, Chamanculo, July 2013.

11 As published in the weekly newspaper *Savana*, 3 September 2010 (our translation).

12 Interview, Ferroviário, August 2013.

13 This is also the line of argumentation developed by Bertelsen (2014).

14 Armando Guebuza, "The Government is aware of the situation in which our marvellous people is living" (our translation), http://armandoguebuza.blogspot.com/2010_09_01_archive.html.

15 Speech by José Pacheco reported in *O País*, 2/09/2010. 'Bandits' is also a term traditionally used by Frelimo leaders to describe members of the armed opposition group Renamo.

16 Article 4 of Ministerial Diploma no. 153/2010, Boletim da República, 1st Series, no. 37, 15/09/2010.

17 The deadline for registering mobile phone numbers was extended from November 2010 to January 2013, after the operators complained that the period was too short for it to be feasible to register some 7 million mobile phone users.

18 President of the Republic, Armando Guebuza, in *O Pais*, 2 September 2010.

19 The Strategic Urban Poverty Reduction Programme, commonly known as "the 7 Million" after the value in local currency of the funds initially allocated to each rural district, promised to expand the coverage of these funds to the cities. Its stated aims were to increase job opportunities, to improve the business environment and the employability of labor and to improve the social protection system.
20 Baptista Macuvele, deputy chairperson of Maputo Association of Road Transporters, interviewed in Maputo, 29 November 2013.
21 Ministry of Industry and Trade, *Manual de Procedimentos Do Subsídio à Cesta Básica*, Maputo, April 2011.
22 PAPA was approved in June 2008 by the Inter-Ministerial Commission on Prices.
23 People are not enthusiastic about cassava bread, they are concerned that the flour is difficult to store and there are also no guarantees that its price could be lower than wheat flour. Despite these uncertainties, the government has been encouraging the production and marketing of cassava while more in-depth studies are under way, but although an increase in production has been recorded in recent years, this is mainly due to increased demand resulting from a venture to brew beer using cassava as raw material that was set up in Nampula in 2012.
24 A. Muianga, National Director of Trade, Interview, 4/12/2013, Maputo.
25 Among the package of measures, there stood out: 1) the increase in the Bank of Mozambique's interest rates for intervening in the money markets, namely the Standing Lending Facility and the Standing Deposit Facility, both by 100 base points, bringing them to 16.5 per cent and 5 per cent respectively, and raising the Compulsory Reserves Coefficient by 24 base points to 9.0 per cent, as from the period of constitution that began on 7 February; 2) intervention on the Inter-Bank Exchange Market, providing foreign exchange to the market to a total of 162 million USD from the beginning of the year to 23 February. Of this sum, 57.1 million USD was destined to cover January's fuel invoice.
26 Of the manifestos prepared for the most recent general elections (2014), only Frelimo's explicitly mentioned food security, while the issue was addressed only implicitly in the MDM manifesto and was completely absent from Renamo's.

References

Arndt, C., S. Jones, and V. Salvucci. 2015. "When Do Relative Prices Matter for Measuring Income Inequality? The Case of Food Prices in Mozambique." *Journal of Economic Inequality* 13(3):449–464.

Bertelsen, B. E. (2014). "Effervescence and Ephemerality: Popular Urban Uprisings in Mozambique." *Ethnos: Journal of Anthropology* 81 (1): 25–52 http://dx.doi.org/10.1080/00141844.2014.929596

Bertelsen, B. E. 2016. *Violent Becomings: State Formation, Sociality and Power in Mozambique*. Oxford: Berghahn.

Bohstedt, J. 2016. "Food Riots and the Politics of Provisions From Early Modern Europe and China to the Food Crisis of 2008." *The Journal of Peasant Studies* 43(5):1035–1067.

CIP. 2010. "Sobre o Registo de Cartões SIM : Diploma Ministerial incoerente, ilegal e anti-Constitucional." Observatorio de Direito n°1. Available at http://cipmoz.org/images/Documentos/Anti-Corrupcao/54_Observatório_de_Direito_n°_1.Documento_do_CIP_2010.pdf

Chaimite, E. 2014. "Emergência de um Repertório de Acção Colectiva em Moçambique." In *Desafios para Moçambique – 2014*, edited by Luís Brito, Carlos Nuno Castel-Branco, Sérgio Chichava, and António Francisco. Maputo: IESE.

Chen, Jidong, Jennifer Pan, and Yiqing Xu. 2016. "Sources of Authoritarian Responsiveness: A Field Experiment in China." *American Journal of Political Science* 60(2):383–400.

Froissart, C. 2014. "The Ambiguities Between Contention and Political Participation: A Study of Civil Society Development in Authoritarian Regimes." *Journal of Civil Society* 10 (3):219–222.

Governo de Moçambique (2011), *Pacote de Medidas para Atenual o Custo de Vida Após o 1° Trimestre de 2011*, *Apresentação* ao Conselho de Ministros, Maputo, 29 de Março.

Hanlon, J. 2016. "Mozambique Inequality Doubles in Six Years, According to Survey." *Mozambique* 306, 4 January.

Honwana, A. 2013. *O Tempo da Juventude: Emprego, Política e Mudanças Sociais em Africa*. Maputo: Kapicua.

Hossain, N., and D. Kalita. 2014. "Moral Economy in a Global Era: The Politics of Provisions During Contemporary Food Price Spikes." *The Journal of Peasant Studies* 41 (5):815–831.

Mandlate, O. 2015. "Vulnerabilidade dos estratos urbanos pobres: o caso da pobreza alimentar em Maputo." *Boletim IDeIAS*, 80. Maputo: IESE.

Mosca, J. 2014. "Produção Alimentar: Um problema central por resolver." *Destaque Rural*, n° 3. Maputo, OMR. www.omrmz.org/images/publicacoes/D3.pdf

Pereira, Cresêncio, Egídio Chaimie, Lúcio Posse, and Michael Sambo. 2013. As "revoltas do pão" de 2008 e 2010 na imprensa. *Boletim IDeIAS*, 58. Maputo: IESE. www.iese.ac.mz/lib/publication/outras/ideias/ideias_58.pdf

SETSAN 2013. *Informação Sobre a Implementação da Estratégia e Plano de Acção de Segurança Alimentar e Nutricional (ESAN II)*. Maputo: SETSAN.

Tvedten, Inge, Sandra Roque, and Bjørn Enge Bertelsen. 2013. "Espaço Urbano e Pobreza em Maputo, Moçambique." *CMI Resumo*12 (2). www.cmi.no/publications/file/4794-espaco-urbano-e-pobreza-em-maputo-mocambique.pdf

Van den Boom, B. 2011. *Análise da pobreza em Moçambique – Situação da pobreza dos agregados familiares, malnutrição infantil e outros Indicadores – 1997, 2003, 2009*, www.sow.vu.nl/pdf/Mozambique/Analysis%20of%20Poverty%20in%20Moz%20March%202011%20Port.pdf

World Bank. 2008. *Mozambique - Higher fuel and food price: impacts and responses for Mozambique*. Washington, DC: World Bank. http://documents.worldbank.org/curated/en/640711468287778861/Mozambique-Higher-fuel-and-food-prices-impacts-and-responses-for-Mozambique

Wuyts, M. 2011. Does Economic Growth Always Reduce Poverty? Reflections on the Mozambican Experience. *Boletim IDeIAS*, 35E. Maputo: IESE. www.iese.ac.mz/lib/publication/outras/ideias/ideias_35E.pdf

9 How 'food riots' work, and why they matter for development

Naomi Hossain and Patta Scott-Villiers

Introduction

In this concluding chapter, we draw together the threads and themes of the arguments and evidence presented in this volume to arrive at two main conclusions. The first is a conclusion about how food riots worked, by which we mean how subsistence protests regularly served as critical junctures in political struggles over recent food price crises. In order to understand why such events created the conditions for changes in the political and policy conditions for food security, we follow McAdam, Tarrow and Tilly (2001) in dissecting these events until we properly understand their constituent elements. We then build them back up till we understand how these mechanisms interacted in the contentious politics of provisions of the recent global food price crisis. The deconstruction of these events allows us to identify common or recurring elements – particular ideas about food or the responsibilities of public authorities, highly specific market conditions or political or policy responses often found around 'food riots'. It helps to understand what features of these events are rightly understood as the natural or biophysical response of humans faced with subsistence stress, and what, by contrast, is the properly political behaviour of strategic, knowing agents disturbing the balance of power for the purposes of retaining or increasing their subsistence rights.

Deconstructed into their most commonly recurring mechanisms, we start to see that a reason why food riots have an apparently universal resonance is not that they are a physiological response to basic human need, but because such need is particularly successfully politicized. This appears to be especially true when political histories of dearth are part of public memory, and the eradication of dearth is part of the current mandate of the state. Food riots dramatize the moral priority of human eating over other laws or rules, challenging in particular the right to profit from speculation in the foodgrain trade. Our findings suggest that in those states that are most effective at addressing food crises, the rulers often substantially agree with those they rule on a key political principle: that a legitimate and effective state is one that at the least tries to protect its population against major subsistence crises.

When we break subsistence protests down in this way, then, it is evident that the reconstructed contentious politics of provisions are, in the first instance, distinctly national struggles. They lock the rulers of nation-states and their populations into negotiations that relate each to the other as citizens and their states, in struggles over rights and duties, the means of provisioning and the substance and legitimation of public authority. The politics of provisions start to reveal themselves as substantive matters of state-citizen relationships, and of processes through which the state and its citizens come to know each other, to understand each other's capacities and needs, and to find ways of shaping the behaviours and responses of each. Subsistence protests then stop appearing as the peripheral or momentary rumbles of the belly, and are shown to be more thoroughly constitutive of core processes of socio-political and economic development in contemporary developing countries, just as they were for the state-citizen relation in the development of states in the now-advanced economies (C. Tilly 1975).

This brings us to the second main conclusion of the volume, which is about why such events matter for development. We argue here that food riots are critical junctures in political negotiations over food security, and that their prevalence and the form they take – whether wage strikes, 'ration riots' or bargaining over the rightful operation of food markets – closely reflect the type and degree of not only social and economic development, as Louise Tilly has argued (1983), but also of political development (Moore 2001). By this we mean that the means by which a state resolves the politics of provisions – how it structures the communication of fundamental discontent, and its capacities to respond in a broadly satisfactory and socioeconomically developmental manner – is a powerful determinant of state formation more generally. What a state does to ensure that its population – particularly its vocal, mobilized population – can be adequately fed determine whether it can protect national security, develop a functioning administrative capacity and accumulate the surpluses that will permit transition from agrarian to industrial economies.

A focus on food riots draws attention to how and whether the state-citizen relationship has developed the capacities to anticipate, listen and respond to mass food security concerns, and is therefore authoritative, effective and accountable with respect to foundational problems of food insecurity. But the globalized nature of the world food system and of recent food riots raises questions about whether such capacities can realistically be nation-centered under conditions where economic development is inherently a matter of global integration. Such questions are particularly important as we face the immediate future challenge of climate change and its effects on world food security.

Some definitions

'Food riots' as rights-claiming protests

It is worth revisiting what we mean when we talk about 'food riots' now that we have examined evidence of the set of recent subsistence protests, and five country cases in closer detail. By 'food riots', as we noted in the Introduction (Chapter 1, this volume), we have been referring to 'unruly actions of the crowd to assert a right to food'. These are contentious political events that move beyond the formal or customarily legitimate modes of political engagement by the masses, and in both the content of their claims and the manner of their claims-making, challenge the interests of powerful groups with respect to how food is marketed. 'The crowd' refers to non-elite groups – specifically here, groups with grievances about the food system. The manner of non-elite group assertions about the right to food may in practice take a wide range of forms, from the polite petitioning of organized civil society to the tire-burning of the apparently spontaneous Mozambican street battles of 2010. But the focus on unruly, or rule-breaking but ultimately state-rectifying, acts reflects that there is something specifically important about protests that escalate into social and then into political unrest. In other words, there are particular reasons why such types of transgression work to disrupt power relations around the food supply and to dislodge, if often only momentarily, policies that serve the interests of powerful groups, and why, therefore, our attention is drawn to such unruly acts, over and again, across world history and geography.

It is also important to clarify that when we define food riots as a process of claiming the right to food we are speaking of that right in a broad or 'common' sense, rather than as specifically in relation to that right as established under international law (Hossain, te Lintelo, and Wanjiku Kelbert 2015). It is true that the riots and protests illustrated in this volume called for a specific contemporary version of the right to food: the right to be protected against undue volatilities in the price of food, in a market-based food system. They assert a right to eat over a right to profit from food markets. But this claim to protection against markets is not, in the first instance, derived from a utilitarian calculus over the best means of legislating between different claims to maximize fairness so much as it is derived from the ethical principles of a natural right – in extremis, the right to *be*, "possessed simply by virtue of being a human being" (Donnelly 1982: 397). We are therefore referring to the right to food as "primarily a moral claim as to what should be valued, and what institutional structure we should aim for, and try to guarantee if feasible" (Sen 1988: 64; see also Sen 2004). We believe – and we believe that people in general believe – that if there is a claimable right to protection against shocks in the food system, it derives from the natural right for humans to be. This resort to natural rights as the source of the right to food gives it priority as a principle over other rights under the law; for instance, property rights. It also implies an adjudication between competing rights (Sen 1988; also Zuckert 1996 on the relationship between natural law and natural rights).

In our present empirical analysis of claims-making in relation to the right to food, we are directed to the natural status of the right to food as an empirical matter of how those at risk of hunger view the matter. The arguments of protestors frequently include reference to natural rights and natural law in the idea that unruly protest (law-breaking, often) and the assumption of rights over food commodities (or property – taking over markets, looting, price-setting, blockades) is justified by a higher principle of the right to survival, or the right to be. That said, protestors also articulate legal and instrumental bases for such claims, in contexts where a right to food has been established in law and/or where the smooth supply of food is presented as essential to the wider problem of a well-ordered society.[1]

'Food riots' as critical junctures

The right to food as manifested in claims to protection against the failures of food markets to enable secure access to food is articulated in what has been termed the 'moral economy' in social historical studies of food riots in the periods of adjustment to market economies in 17th and 18th century Europe (Thompson 1971, 1991; Bohstedt 2010, 1992). There are many versions and views of the moral economy, but we take a comparatively narrow approach to the concept which is close to its original formulation in social history. We agree with Palomera and Vetta (2016) that the term has most power not as a generalized account of social values – that people value non-commodified relations in some spheres of life – but as a concept of the popular political response to changing class relations in (emerging) capitalist economies. We narrow it further by using the term with reference specifically to the politics of subsistence. The idea of a moral economy may inhere in a latent way in popular political ideologies of the rights of people and the responsibilities of their rulers more generally; whether, and the extent to which it does, is an empirical matter. But in our view, the moral economy is most visible and most programmatic at moments when markets substantially fail to assure basic subsistence, and when collective action articulates grievances about such market and regulatory failures. They resonate particularly loudly when arrangements for social reproduction are in transition towards greater commodification. With reference to the period 2007–2012, we treat the concept of the moral economy as helpful in understanding social and political perspectives on how food markets ought to work (Hossain and Kalita 2014). Our use of the idea of the moral economy enables us to access:

> the political culture, norms and expectations of the various groups of people involved in social reproduction, broadly speaking; the power relations between the governed and the elites; and the articulation of such dispositions and relations with capitalist processes of continuity and change.
>
> (Palomera and Vetta 2016: 3)

Expressions of the moral economy are a political critique of how emerging market relations in food systems disempower the consumer masses in favor of market actors such as food corporations, grain dealers and traders and (to a lesser degree) of retailers. It is not, our evidence suggests, a critique of smallholder or medium-sized food producers so much as a challenge to the moral rightness of profit-making through the commodifying effects of trade and speculation.

The transgressive political actions on which we focus here legitimate their claims with reference to the moral economy – specifically, a shared understanding that food markets ought to work to enable people to eat at a standard they believe to be consistent with their custom and status. But by no means do these specific, short-lived events, often in a brief and localized series of identifiably connected protests, tell us the entire story of the politics of provisions. But by rupturing politics as usual, these individual moments – spikes of violence – on which we centre our analysis in this book, indicate the core issues of power and material struggle that are in contention in that place at that time. These units of analysis build up into bigger episodes and processes, many of which combine in wider struggles over the responsibilities of states and the rights of markets and people, which together comprise the politics of provisions. So although we talk a great deal about food riots, because they give us something to grapple with, it is not only the spikes we are interested in. We are interested in making sense of them in light of the negotiations, and the material and institutional positions at stake in bringing about a system that is functional for human wellbeing at its most natural level.

How food riots 'work': Identifying mechanisms

How do these different elements fit together in wider sequences or recurrent patterns that make a difference to the subsistence settlement? It is only possible to unearth this when we consider the wider cycles of contention and relationships between protestors and public authorities in the food system. McAdam, Tarrow and Tilly (2001) make a persuasive case that by studying how individual events shift, confirm or widen the focus of contention, connecting protestors of similar profiles with different or new articulations of the grievance, it is possible to see the longer socio-political changes relating to a right to food in the making. To do this it is important to uncover the mechanisms that more or less predictably work to bring about such change, and how these sequence and interact, within ongoing processes of social and economic adjustment (such as urbanization and de-agrarianization). The critical global event that was the sudden and sustained rise in staple food costs around the world in 2007–2008 accelerated the pace and drama of economic adjustment, with the usual bargaining over rights between citizens and state becoming especially fraught.

Following McAdam, Tarrow and Tilly (2001), we classify the mechanisms we identify from the case studies into environmental, relational and cognitive types of mechanisms. It is already clear that there are similarities between most of the food-related protests across the otherwise different settings we have included in this book. Sharp rises in the price of staple foods were a feature of the social and economic processes within which most of these protests took place. Chapter 2 of this volume confirms, as other studies (Berazneva and Lee 2013; Brinkman and Hendrix 2011; Hendrix and Haggard 2015; Bellemare 2015; Arezki and Brückner 2011; Smith 2014) have shown, that food price rises during 2007–2012 were largely sudden and often unprecedented; while the concentration of protests around periods when prices peaked is in itself suggestive of food riots 'working' to force price controls or other actions to bring food costs down. The significance of food price spikes is also suggested by the counter-example of India, where food prices rose more gradually and were not subject to spikes emanating from the global system (Chapter 6, this volume). Here, food-related protests focused on the implementation of specific policies relating to food provision, and were part of wider political struggles at the national and state levels, rather than focusing on food prices.

Protestor connections

Food price rises may in the first instance have been experienced by individuals, and their families. But by definition, protests are group actions. What united and enabled the mobilization of distinct groups? Why did some groups then take up protest actions, while others did not? Our cases suggest, as others have also argued (in particular, Bohstedt 1988), that protestors were already connected to each other, whether by occupation, proximity or through histories of community politics (see also the more contemporary cases in Auyero and Moran 2007). In Madhya Pradesh, for example, political events claiming the right to food were led by the *Adivasi Adhikar Manch*, an organization of indigenous people with at least a decade of mobilization over education provision, access to agricultural land and fair prices for forest produce. Protests about food prices sometimes built on pre-existing protest groups or organizations, as happened in Kenya when citizen movements 'hijacked' national events with colourful protests. In Maputo in Mozambique, protests started in the areas around transport hubs, where commuters were suddenly faced with sharp rises in fares to places of work, while garments workers near Dhaka in Bangladesh came together through their common places of work, often becoming mobilized by activists spreading the word on public transport.

An environmental mechanism of considerable and recurrent significance is the amplification of protestors' grievances, in some cases facilitating the diffusion of protests to other settings. Social media was a factor enabling the expansion (or 'scale-shift', as McAdam *et al.* [2001] put it) of localized food riots into a national movement in Cameroon (Chapter 5, this volume). In Kenya, innovative, theatrical protests eventually attracted the attention of the media, enabling the Bunge La Mwananchi group to reach a national audience with their protests. In all of these countries, the international media covered some of these protests, in particular those where violence and political unrest triggered state repression; this was particularly important in Bangladesh, Cameroon and Mozambique.

Behaviour of authorities

Another environmental mechanism that recurs in several instances is the local presence of a state actor with responsibilities for realizing the right to food. In West Bengal in India and in Ikutha and Mathare in Kenya, protests centred around public agencies or their contractors with responsibilities for delivering subsidized or free food. In each instance, protestors detected corruption and unfairness in the distribution of relief food, and sought to highlight this food injustice, or to reallocate it themselves. This mechanism highlights the importance of physical proximity to meaningful objects of protest: there were no protests, to our knowledge, against the global food system, although price rises were widely triggered by conditions at that level. Policy responses to the food price spike or protests also introduced new opportunities or threats, feeding into the amplification or diffusion of riot events, or new cycles of repression and response. This is in sharp contrast to the abstract, de-contextualized nature of the source of the global food crisis. It highlights the problems of scale in popular resistance to the downsides of globalization (Fraser 2012, 2011; Engels 2014) and explains why protesters focused their attentions strongly on the state, even when they were aware of the global forces that were at play.

Food-related protests were also connected with several shifts in relationships between protestors, states, markets and wider society. A key relational change was in how protestors sometimes sought to naturalize their concerns with reference to basic human needs and rights. The language of the moral economy made it possible for particular groups to articulate their own concerns and interests with reference to those of the wider society, drawing sympathy and/or increasingly the likelihood that their protests would be 'certified' as legitimate protests. The garments workers of Bangladesh are a good example of this. Although their struggles over low wages pre- and post-dated the 2008 food price shock, they articulated their concerns at that particular moment with reference to high food prices and the right to food, with slogans like 'bring prices down, let us survive'.

Powerful markets

Food trade interests were regularly demonized, including both oligopolies or powerful groups with the ear of policymakers, and more tangibly and more visibly to food shoppers and protestors, food or grain traders or ration-dealers. Those who profited from food trade – but had not grown it themselves – were often framed as exploitative. This created a target for and helped legitimate protests. In Kenya, the maize flour millers and brokers were seen to have benefited from the government's maize flour subsidy, supposedly intended to benefit the poor, entering "the lexicon of Kenyan scams in the media, on social media and the word on the street in the informal settlements" (Chapter 7, this volume). In an important sense, the demonization of food traders can be seen as an attempt to 'de-certify' their interests and concerns as legitimate in the politics of provisions.

Food price spikes produce a particularly rapid shift in what Araghi (2003) calls the value relations between labor and subsistence, moving against the interests of worker-consumers. Unless prices fairly rapidly drop back to 'normal' levels, wages quickly become inadequate. This means discontent spreads from what people are eating to the nature of the work they do, and the remuneration they earn. Other research into the effects of food price rises after 2012 in which the editors of this volume have been involved found increased precariousness in the kind of labor many people on low incomes were performing, as they found it impossible to make ends meet without doing more to earn more (Scott-Villiers *et al.* 2016).

Speaking about how the minimum wage had adjusted to the price rises, an official from the Mozambican Workers' Organization (OTM) noted that:

> The evolution of the minimum wages is not satisfactory, because it is still not enough to acquire the desirable basket of goods for a family of five people. We see that the government does not take responsibility for protecting the producers of wealth. When the government presents us with data on jobs created, we ask "what kind of jobs?" What we want is to abandon the negotiation of wages on the basis of percentages, and move to negotiating on the basis of the basic basket. But the government insists on negotiating based on the average inflation and on the growth in GDP per sector. How many citizens are able to buy milk? How many kilos of rice are citizens able to buy? The government does not have an employment policy. Today jobs are allocated personally, and it's full of corruption.
>
> (de Brito *et al.* 2014, 15)

In articulating their grievances about price rises, labor organizers seek to challenge the delinking of labor from a reasonable living, and in so doing highlight the generally inadequate nature of the labor market and its regulations and provisions, as was the case in Bangladesh. This shift in the value relations between labor and subsistence is a framing perhaps more latent than present in the other food riots we have considered here, but it may become a mechanism when this shifting value-relation is collectively attributed as a threat to the interests of labor (on the mechanism "threat attribution", see McAdam, Tarrow, and Tilly [2001]; Chapter 4, this volume).

The effect of food riots

States and rights

A strong emphasis on subsistence protection, often as a result of the damaging political effects of past food crises or disasters, can create a relationship of accountability between citizens and the state in relation to protection against food shocks. This is true in both India and Bangladesh, with major famines at the heart of their national liberation struggles. But the mechanism works in different ways, depending on the institutional response to such accountability for food security, and on whether and how protestors are positioned to demand answers from the state. A final relational mechanism of great importance is brokerage, or the forging of new alliances that enable weaker or less organized groups to access the organizational resources of stronger ones. Most of the cases have instances in which protest groups were brokered into more effective or more powerful networks through connections made with external actors.

The idea that there is a 'common sense' (as distinct from a legal) right to food – and that sudden, high food price rises are unjust – can give rise to views that under such conditions, the state is responsible for action. This perception means that if and when a political opportunity arises (a threat or opportunity is attributed to some action or inaction by the state), the state becomes the target of protests or claims relating to access to food. We have seen that elites, media and masses all tend to believe that food price rises lead to discontent, and expect food price rises to lead to food riots and political danger. This is a cognitive mechanism of great significance, because it highlights the dynamic and circular nature of food riots, and the connections between protests in one period, protests in another, and how the behaviours of elites and protestors anticipate the responses of the other based on historical memory of subsistence crises past. We also know from the media analysis in Chapter 3 of this volume that elites typically expect diffusion of food riots from one place to another, and fear the unpredictability that they bring to the political situation, and may therefore be inclined to repress protest. This is always a risky strategy, it may deter protestors, but as in the Cameroon case, violence against protestors perceived to have legitimate concerns can escalate the situation as protestors protest their right to make their claims.

How do the different elements combine in processes of political change? The country studies demonstrate the influence of local histories of subsistence politics, conjoined political and economic crisis, and past success with popular mobilization around food; these can mix in combustible combinations in the political imagination when food prices spike too suddenly for 'natural' events to explain. These wider processes of food price struggles may be triggered specifically by food price spikes affecting groups that are already mobilizing around their rights, economic and political. With McAdam, Tarrow and Tilly (2001), we find that static, single-shot analyses of food riots fail to capture the lessons of history and their institutionalization in policies, programs, moral economic ideas and beliefs, as well as expectations about the behaviour of the different stakeholders.

Figure 9.1 illustrates the previous discussion. Building on McAdam, Tarrow and Tilly's (2001) "dynamic interactive framework for analysing mobilization in contentious politics", we apply their mechanism-based explanation to contemporary food riots. The framework shows that food price rises are a key part of the wider social and economic processes within which food riots take place, but do not in any sense *cause* them. A mixture of the connections between those under stress, the behavior of authorities and market actors and beliefs about a moral economy combine to cause riots. In particular, where low income populations have been previously mobilized, live in proximity or share common occupations, it becomes easier for the collective attribution of threats and opportunities arising from food price rises and policy responses to them. Where there is a strong belief, based on past experience, that the state has reasons to take its responsibilities to protect citizens against food crises seriously, and in particular where that has translated into the establishment of local food grain or ration dealers or officials tasked with realizing the right to food, protests may be 'certified' or rendered legitimate. States do not change because of riots, but are already always changing with global economic adjustment; riots and protests are part of that change. Based on past experience, elites (global and national) expect high food prices to generate discontent, and even riots; they fear such outcomes because

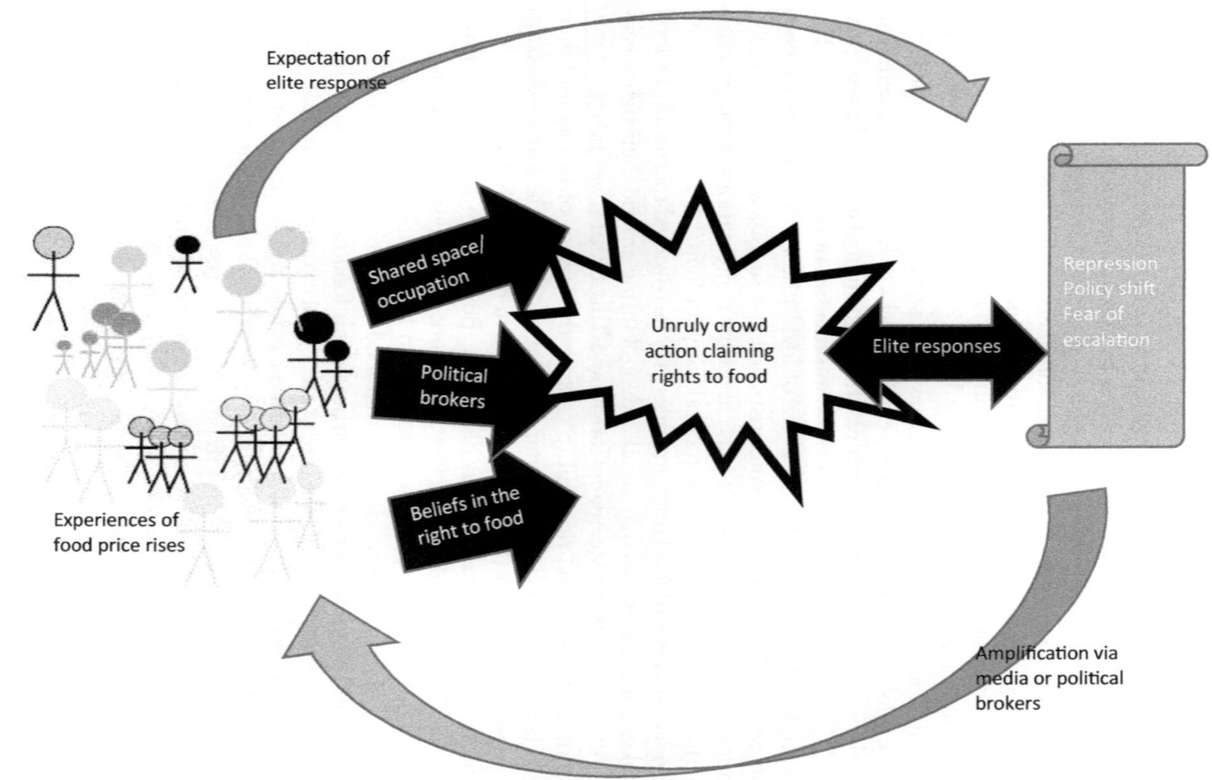

Figure 9.1 A dynamic framework for analyzing contemporary food riots

('legitimate') protests about food are perceived to easily escalate, particularly if and when the media amplifies the voices of protestors. Elite fears can lead to violent repression, which may further ramp up protests as rioters claim their rights to claim their rights at all. Other movements or civil or political society actors (trades unions, political parties, NGOs) may step in and consolidate the protestors' platform, building coalitions of greater strength and capacity, and potential threats to incumbent powers. A state that responds without repression but with price controls or subsidies reinforces the perception of state responsibility.

Do food riots indicate development failures?

We noted previously that our interest in food riots arose from the insights these events offered into the larger process of negotiations over food security between citizens and states. The prevalence and nature of such events (who protests, how frequently, about what, when and how, and who they target) both respond to and create the conditions for food policies and natural rights-bearing citizenship. This political sociological perspective on the significance of food riots for processes of socioeconomic and political development follows Charles Tilly (1975) in emphasizing state-citizen relations and state capacity to enable food security, through the political capacities to anticipate, listen and respond to popular or moral economic concerns, but also the policy and infrastructural capacities necessary to integrate and regulate food markets, including information and communication networks.

Our key departure from other contemporary food riots analyses is that we do not see such protests as signs of dysfunction and systemic failure so much as evidence of bargaining and negotiation during periods of structurally complex and difficult adjustment. They are, therefore, a variety of political struggle that is entirely necessary for states to learn how to anticipate, listen and respond to popular concerns about subsistence crises. By contrast, both major schools of thought on recent food riots, the development policy perspective and the food regimes view, see these as evidence of systemic failure. The question for us, then, is whether this bargaining can operate effectively within the increasingly complex and increasingly globalized arena in which the politics of provisions must operate in the 21st century.

Security failures

In the development policy literature, recent food riots are a clear signal of momentary but major market failure. Development policy research is primarily concerned with food riots as social or civil unrest, with the potential to escalate further into violence and declines of law and order (Arezki and Brückner 2011; see also Chapter 3 in this volume, on responses of Bretton Woods institutional elites to food riots). Drawing on a half century of data from Sub-Saharan Africa, Hendrix and Brinkman (2013) characterize the relationship between food security and conflict as circular, viewing food security as a "threat multiplier" (see also

O'Brien 2012). There is an understandable emphasis on the need to prevent such political violence: people die in food riots, mostly from official repression; livelihoods that rely on the food trade may suffer, and property may be damaged. As we know from the classic Walton and Seddon (2008) account of the 'IMF riots' of the 1970s and 1980s, liberalizing economic policies may be derailed by political elites keen to regain popular support in the wake of mass protests against the effects of economic adjustment.

Although prices are key variables, most development policy scholarship treats food riots and related protests as political acts, within the constraints of the data on which it typically relies. Berazneva and Lee, for instance, attempt to explore the extent to which 'African food riots were a response to limited food availability or to restricted access to food, or both' (Berazneva and Lee 2013: 32). This follows Amartya Sen in shifting attention from the Food Availability Decline thesis of famine to a political-relational view of food insecurity as resulting from the erosion or abrogation of food entitlements (Sen 1981b, 1981a). They conclude that although food availability, levels of poverty, features of local food markets and urban and coastal location all helped determine the prevalence of food riots, more oppressive regimes and stronger civil societies also, somewhat paradoxically, meant food riots were more likely to occur (Berazneva and Lee 2013). They conclude that food riots are political signals that need to be addressed through politics and policies that redistribute power, and not just power over food:

> African governments need to effectively deal with food access and distribution problems to avoid future food riots. By helping poor urban and rural households maintain access to food, energy and essential services, social protection programs . . . can reduce the impact of price shocks on health and education, as well as prevent social unrest . . . Dissatisfaction with government policies and corruption, as well as desires for increased freedoms, were among the major complaints voiced by the public . . . Such discontent with matters of governance and political rights is not surprising in Africa.
>
> (Berazneva and Lee 2013: 37)

Hendrix and Haggard take the political analysis further, testing a series of hypotheses about the political-institutional determinants of food riots on a dataset of urban Asian and African unrest since the 1960s. They conclude that (cf. Berazneva and Lee 2013) democracies and semi-democratic polities have seen more protests when food prices rise for two main reasons: because the political opportunities are greater than in more authoritarian regimes, and because more democratic polities tend to feature relatively less urban bias, and pro-farmer policies often mean higher prices, particularly for low-income urban populations (Hendrix and Haggard 2015). Analyses of this kind provide valuable frameworks for the analysis of what food riots mean for development, and Hendrix and Haggard and Bellemare are among scholars that call for deeper engagement with the dynamics of protests and protestor motivations of the kind we have aimed for here.

Food regime failures

The main alternative perspective on contemporary food riots is from food regimes thinking. This approach also treats food riots as evidence of failure, but for scholars like Raj Patel, Philip McMichael and Ray Bush, this failure is more systemic and total, from the point of view of protestors, rather than the politically-inflected protests against market failures of development policy research. As discussed in the Introduction (Chapter 1, this volume), food regime theory offers a framework for understanding changes in the world food system in relation to how agriculture is integrated into and serves global processes of capital accumulation (Friedmann 1980; Bernstein 2016; McMichael 2009a). The state of food security in any particular time and place reflects how it is incorporated into the regime. For food regimes theorists, the so-called global food crisis of the early 21st century was 'crisis as normal' (for instance, Lang 2010; McMichael 2009b). In other words, the world food system was broken to the extent that the orientation of the current food regime towards serving the interests of corporate capital was no longer affording even a viable basic existence for labor.

For Bush, mass unrest relating to the food system is intrinsic to the gross injustices and inequities of that system:

> [F]ood insecurity and famine is an integral part of late capitalism. There should no longer be surprise that death by starvation and chronic malnutrition are systemic features of capitalism and that people resist.
>
> (Bush 2010: 119)

Food riots are "just one acute form of structural, historical patterns of underdevelopment that are shaped by and in turn shape the political economy of food" (Bush and Martiniello 2017: 193). Struggles over the food system are necessarily violent struggles over bare survival: for instance, for McMichael, the global wave of food riots and related protests was evidence that 'agflation is generating elemental struggles around the question of social reproduction' (McMichael 2009b, 285–286). As with the development policy approach, food rioters are not spasmodic in their protest; although elemental struggles, these are also properly political responses, even if (in contrast to the development policy research) the response is to the deepening dependency on food trade rather than to its momentary failure. For instance, the 2008 wave of food riots was:

> an outcome of the policies embodied in the Bretton Woods Institutions' economic doctrine, insofar as they dismantled public capacity (specifically food reserves), and deepened food dependency across much of the global South through the liberalization of trade in foodstuffs.
>
> (Patel and McMichael 2009: 10)

Food regime theory provides a robust framework for examining the significance of food riots for development, showing that the food riot is 'not simply about the price and accessibility of staple foods [but about the more complex matter of] the political economy of food provisioning' (Patel and McMichael 2009, 12). 'Food provisioning', Patel and McMichael go on to argue, 'serves as a lightning rod of contention over perceptions of just governance' (pp. 24). 'Food rebellions' are important as evidence of significant popular alternatives to integration into the global food system, and in particular signs of mass support for more sustainable, agro-ecological and less commodified food systems; the international food sovereignty movement is evidence of this politics of alternative food provisioning (Holt-Giménez and Patel 2009).

Food sovereignty and its transnational organizing may well provide the ideological alternative to a fully commodified global food system. But how and the extent to which nation-states are capable of establishing and protecting their own alternative food systems at least partially protected from the vicissitudes of the global economy will depend substantially on state capacities, as noted previously, to anticipate, listen and respond to popular concerns about the food supply. We recognize that the politics of provisions are about negotiations between states and citizens, and that the room for maneuvering of states facing contemporary food rioters is different to that faced by the British and French states in their own transitions to capitalism. The experiences of historical processes of state formation in now-developed countries are not, of course, identical or even broadly similar to their former colonies, the 'developing' economies of the present. There are many reasons for this, but the one that matters most directly in the discussion of the historical development of state capacities to manage food security is the comparatively globalized nature of many contemporary developing countries' food systems. As the Indian case study in this volume notes, a significant difference between India and the other country cases is its comparative autarky and capacity to make, rather than take, global (or at least regional) food prices. For the other countries in the set, integration into – and vulnerability to contagion from – different segments of the global food system was an important part of the environment within which the politics of provisions erupted in unrest during the 2007–2012 period.

Globalizing times and the politics of being

The global wave of recent subsistence protests encourages us to view these struggles and the international policy responses they engendered as the negotiations of a global politics of provisions. The wave of 'food riots' in the period 2007–2012 reflects the incorporation of the universal everyday practice of eating in the early 21st century within an internationally integrated and interrelated system of food markets, policies, production and regulatory systems. When global food prices spike or slump, this affects almost everyone who eats – but it does so unevenly, depending on how populations have previously negotiated their rights to food, subsistence and, ultimately, status with their public authorities. Whether or not they feel they need to or can usefully protest to reassert those rights in a sharply shifting food market depends on how successfully those negotiations have tilted the balance of power towards those who eat as opposed to those who profit from food.

We have used the term 'right to food' and 'food rights' in an apparently casual and footloose manner, with little to no reference to constitutions or international law, other than in the interesting, contrasting cases of India and Kenya. But we hope it is clear that we see in this global wave of subsistence protests a politics of the right to food – even if such terms are not widely used, other than as specific expressions of international human rights law. But we do not only see this wave of subsistence protests as a global politics of provisions that has returned attention to the food insecurity embedded in a globalized, commodified world food order. We also recognize in the responses of the relevant institutions of global governance (such as they are) an acknowledgement of the right to food, even despite the colonization by the global economy of food systems everywhere and for everyone. The comparatively strong emphasis on agriculture, food security, nutrition and social protection in aid and development policy, and the relatively weak defence of the freedom of markets in food commodity trade is, we believe, an indication of the success of this multi-sited global struggle over the right to food.

A key point of difference in the view of food riots as strategic acts in a political drama of significance for the shaping of contemporary food systems is our view that such protests are productive political negotiations, and not signals of economic or political failure. There is a risk here that by kicking out the hydraulic or spasmodic view of food riots we end up instead with an impression that they are highly deliberative political performances, riot as dance theater rather than as angry violence. But we want to steer clear of seeing food riots as staged events, with the suggestion of fakery; we know that protestors are frequently angry when they protest, and fearful that they and their families may face hunger if circumstances do not change. We are arguing instead that this anger is deployed politically and strategically in a spectacle of protest adopted with the tacit knowledge of the forms taken by popular protest over food crisis historically. The crowd is drawn together in the tense times of economic transition by the magnetic qualities of the hard-won knowledge of the collective power of the mob (Rudé 1981) and of the idea (with a capital 'I') that things could be different and better (Badiou 2012).

The reach and effect of subsistence protests is causally and consequentially related to the level and type of political development with respect to food security and citizen-state relations. These are productive protests – part of improving citizen-state relations, however frightening and regressive in the short term. But what positive role can protest make in the 21st century, given growing turbulence in the global economy, the subsistence crises that must result, and the subsequent world discontent that is sure to come? Can protest make a positive contribution given the vastly encompassing effects of dehumanized food regimes and the commodification of life? If the politics of provisions has shown us anything, it is that protest is adaptable to changing circumstances, and yet retains its political power across many contexts. It seems axiomatic that states and citizens that have developed the capacities to anticipate, listen and respond to their nation's food security needs will be those best equipped for the social and economic transitions from agrarian to complex industrial societies to becoming members of a globally negotiated mega-polity. These capacities will reflect the distinctively different political histories of dearth across countries, which in turn reflect in large part the agro-ecological conditions and geographies of different places. They will take different forms – more openness to global and local markets in the case of smaller economies, and big public systems of food and income transfers in larger ones, for instance. And they will also reflect politics proper, depending on which groups are incorporated into the political bargain, and on what terms. In this book we have shown how riots and protests allow unequivocal voices of suffering to be heard, voices which remind us of the meaning of *the natural right to be*, and insist on a politics of provision that simply will not stop provoking the food system at its most local and at its most global.

Note

1 Although this is not the place for a lengthy discussion of the relationship between natural rights, natural law, human rights and the right to food, in developing our understanding of the right to food as an ideological and programmatic claim made by people at risk of hunger, we have drawn on arguments developed by Donnelly (1982) and Zuckert (1996); see also Sen (1988, 2004) on the right to protection from hunger specifically.

References

Araghi, Farshad. 2003. "Food Regimes and the Production of Value: Some Methodological Issues." *The Journal of Peasant Studies* 30 (2):41–70. doi:10.1080/03066150412331311129.

Arezki, Rabah, and Markus Brückner. 2011. *Food Prices and Political Instability*. Washington, DC: International Monetary Fund.

Auyero, Javier, and Timothy Patrick Moran. 2007. "The Dynamics of Collective Violence: Dissecting Food Riots in Contemporary Argentina." *Social Forces* 85 (3):1341–1367.

Badiou, Alain. 2012. *The Rebirth of History: Times of Riots and Uprisings*. Translated by Gregory Elliot. London. New York: Verso.

Bellemare, Marc F. 2015. "Rising Food Prices, Food Price Volatility, and Social Unrest." *American Journal of Agricultural Economics* 97 (1):1–21. doi:10.1093/ajae/aau038.

Berazneva, Julia, and David R. Lee. 2013. "Explaining the African Food Riots of 2007–2008: An Empirical Analysis." *Food Policy* 39:28–39.

Bernstein, Henry. 2016. "Agrarian Political Economy and Modern World Capitalism: The Contributions of Food Regime Analysis." *The Journal of Peasant Studies* 43 (3):611–647. doi:10.1080/03066150.2015.1101456.

Bohstedt, John 1988. "Gender, Household and Community Politics: Women in English Riots 1790–1810." *Past and Present* 120:88–122.

———. 1992. "The Moral Economy and the Discipline of Historical Context." *Journal of Social History* 26 (2):265–284. doi:10.2307/3788421.

———. 2010. *The Politics of Provisions: Food Riots, Moral Economy, and Market Transition in England, C. 1550–1850*. Ashgate: Farnham, Surrey/Burlington VT.

Brinkman, Henk-Jan, and Cullen S. Hendrix. 2011. "Food Insecurity and Violent Conflict: Causes, Consequences, and Addressing the Challenges." Occasional paper. Rome: World Food Program.

Bush, Ray. 2010. "Food Riots: Poverty, Power and Protest." *Journal of Agrarian Change* 10 (1):119–129.

Bush, Ray, and Giuliano Martiniello. 2017. "Food Riots and Protest: Agrarian Modernizations and Structural Crises." *World Development* 91 (March):193–207. doi:10.1016/j.worlddev.2016.10.017.

de Brito, Luis, Egidio Chaimitie, Crescêncio Pereira, Lúcio Posse, Michael Sambo, and Alex Shankland. 2014. "Hunger Revolts and Citizen Strikes: Popular Protests in Mozambique, 2008–2012." Food Riots and Food Rights: The Moral and Political Economy of Accountability for Hunger Project Reports. Brighton, Sussex, England: Institute of Development Studies.

Donnelly, Jack. 1982. "Human Rights as Natural Rights." *Human Rights Quarterly* 4:391.

Engels, Bettina. 2014. "Contentious Politics of Scale: The Global Food Price Crisis and Local Protest in Burkina Faso." *Social Movement Studies* (May):1–15. doi:10.1080/14742837.2014.921148.

Fraser, Nancy. 2011. "Social Exclusion, Global Poverty, and Scales of (In)Justice: Rethinking Law and Poverty in a Globalizing World." *Stellenbosch Law Review* 22:452.

———. 2012. "Can Society Be Commodities All the Way down? Polanyian Reflections on Capitalist Crisis." Fondation Maison des sciences de l'homme working papers series no. 18. Paris: Fondation Maison des sciences de l'homme.

Friedmann, Harriet. 1980. "Household Production and the National Economy: Concepts for the Analysis of Agrarian Formations." *The Journal of Peasant Studies* 7 (2):158–184. doi:10.1080/03066158008438099.

Hendrix, Cullen, and Henk-Jan Brinkman. 2013. "Food Insecurity and Conflict Dynamics: Causal Linkages and Complex Feedbacks." *Stability: International Journal of Security and Development* 2 (2): Art-26.

Hendrix, Cullen S., and Stephan Haggard. 2015. "Global Food Prices, Regime Type, and Urban Unrest in the Developing World." *Journal of Peace Research* 52 (2):143–157. doi:10.1177/0022343314561599.

Holt-Giménez, Eric, and Raj Patel, eds. 2009. *Food Rebellions: Crisis and the Hunger for Justice*. F First Paperback Edition Used edition. Cape Town: Oakland, CA; Boston, MA: Food First Books.

Hossain, Naomi, and Devangana Kalita. 2014. "Moral Economy in a Global Era: The Politics of Provisions during Contemporary Food Price Spikes." *Journal of Peasant Studies* 41 (5): 815–831doi:10.1080/03066150.2014.895328.

Hossain, Naomi, Dolf te Lintelo, and Alexandra Wanjiku Kelbert. 2015. "A Common Sense Approach to the Right to Food." http://opendocs.ids.ac.uk/opendocs/handle/123456789/6434

Lang, Tim. 2010. "Crisis? What Crisis? The Normality of the Current Food Crisis." *Journal of Agrarian Change* 10 (1):87–97. doi:10.1111/j.1471–0366.2009.00250.x.

McAdam, Doug, Sidney Tarrow, and Charles Tilly. 2001. *Dynamics of Contention*. Cambridge: Cambridge University Press.

McMichael, Philip. 2009a. "A Food Regime Genealogy." *Journal of Peasant Studies* 36 (1): 139–169. doi:10.1080/03066150902820354.

———. 2009b. "A Food Regime Analysis of the 'world Food Crisis.'" *Agriculture and Human Values* 26 (4):281–295. doi:10.1007/s10460-009-9218-5.

Moore, Mick. 2001. "Political Underdevelopment: What Causes 'bad Governance.'" *Public Management Review* 3 (3):385–418. doi:10.1080/14616670110050020.

O'Brien, Thomas. 2012. "Food Riots as Representations of Insecurity: Examining the Relationship between Contentious Politics and Human Security." *Conflict, Security & Development* 12 (1):31–49.

Palomera, Jaime, and Theodora Vetta. 2016. "Moral Economy: Rethinking a Radical Concept." *Anthropological Theory*, November. doi:10.1177/1463499616678097.

Patel, Raj, and Philip McMichael. 2009. "A Political Economy of the Food Riot." *Review. Fernand Braudel Center* XXXII (1):9–36.

Rudé, George. 1981. *The Crowd in History: A Study of Popular Disturbances in France and England, 1730–1848*. Lawrence and Wishart London.

Scott-Villiers, Patta, Nick Chisholm, Alexandra Wanjiku Kelbert, and Naomi Hossain. 2016. *Precarious Lives: Food, Work and Care after the Global Food Crisis*. https://opendocs.ids.ac.uk/opendocs/handle/123456789/12190.

Sen, Amartya. 1981a. "Ingredients of Famine Analysis: Availability and Entitlements." *The Quarterly Journal of Economics* 96 (3):433–464.

———. 1981b. *Poverty and Famines: An Essay on Entitlement and Deprivation*. Oxford: Oxford University Press.

———. 1988. "Property and Hunger." *Economics and Philosophy* 4 (1):57. doi:10.1017/S026626710000033X.

———. 2004. "Elements of a Theory of Human Rights." *Philosophy & Public Affairs* 32 (4):315–356. doi:10.1111/j.1088-4963.2004.00017.x.

Smith, T. G. 2014. "Feeding Unrest: Disentangling the Causal Relationship Between Food Price Shocks and Sociopolitical Conflict in Urban Africa." *Journal of Peace Research* 51 (6):679–695. doi:10.1177/0022343314543722.

Thompson, E. P. 1971. "The Moral Economy of the English Crowd in the Eighteenth Century." *Past & Present* 50:76–136.

———. 1991. *Customs in Common*. London: Penguin.

Tilly, Charles. 1975. "Food Supply and Public Order in Modern Europe." In *The Formation of National States in Western Europe*, edited by Charles Tilly, 380–455. Princeton, NJ: Princeton University Press.

Tilly, Louise A. 1983. "Food Entitlement, Famine, and Conflict." *The Journal of Interdisciplinary History* 14 (2):333–349. doi:10.2307/203708.

Walton, John K., and David Seddon. 2008. *Free Markets and Food Riots: The Politics of Global Adjustment*. 2nd ed. Hoboken, NJ: Wiley-Blackwell.

Zuckert, Michael P. 1996. "Do Natural Rights Derive From Natural Law." *Harvard Journal of Law & Public Policy* 20:695.

Index

For Product Safety Concerns and Information please contact our EU representative GPSR@taylorandfrancis.com
Taylor & Francis Verlag GmbH, Kaufingerstraße 24, 80331 München, Germany

www.ingramcontent.com/pod-product-compliance
Ingram Content Group UK Ltd.
Pitfield, Milton Keynes, MK11 3LW, UK
UKHW011010180926
475701UK00004B/664
* 9 7 8 0 3 6 7 3 5 2 1 5 8 *